ENVIRONMENTAL IMPACT STATEMENT DIRECTORY

The National Network of EIS-Related Agencies and Organizations

ENVIRONMENTAL IMPACT STATEMENT DIRECTORY

The National Network of
EIS-Related Agencies and Organizations

Edited by

Marc Landy

Director, Environmental Impact Statement Institute

IFI/PLENUM • NEW YORK-WASHINGTON-LONDON

Library of Congress Cataloging in Publication Data

Landy, Marc, 1950-
Environmental impact statement directory.

Bibliography: p.
Includes index.
1. Environmental impact statements — United States — Directories. I. Title.
TD194.5.L36   333.7'1'02573   80-27909
ISBN 978-1-4684-6125-1

ISBN-13: 978-1-4684-6125-1   e-ISBN-13: 978-1-4684-6123-7
DOI: 10.1007/978-1-4684-6123-7

© 1981 IFI/Plenum Data Company

Softcover reprint of the hardcover 1st edition 1981

A Division of Plenum Publishing Corporation
227 West 17th Street, New York, N.Y. 10011

To KATHI and JOE

PREFACE

This book sets out to provide a useful directory to writers, reviewers, and citizens interested in the environmental impact statement (EIS) process in the United States. Although several "environmental" directories address a narrow portion of EIS subject matter, there is no single directory that covers the wide range of agencies and organizations involved in the EIS process. Therefore, the purpose of this directory is to bring EIS-related agencies and organizations into one comprehensive reference book.

EIS writing and reviewing can be improved by providing practioners with a directory that puts them quickly in touch with the national EIS network. At the present time, practioners must rely on their personal, and sometimes, limited network of agencies and organizations for the exchange of information. The EIS Directory quickly connects EIS-related agencies and organizations throughout the United States at the national, regional, state, and local levels.

The EIS Directory is a compilation of directories provided by members of the EIS community. A written survey of over five hundred agencies and organizations was made in 1980. The response to the survey provided a wide range of in-house directories. This book contains approximately 4,000 entries developed from over 300 documents.

This book was written in the hope that it will help people communicate more easily--ultimately improving the quality of EIS decision making. I realize the limitations of this initial edition and accept the challenge of providing a more effective reference tool in future editions.

Many people have contributed their time and energy to this project. My thanks go to Rick Thacker and Oscar Graham for reviewing portions of the manuscript. Plenum Publishing Corporation has been most helpful with particular appreciation to Frank Columbus, Director of Product Development/Editor; John Matzka, Managing Editor; and Georgia Prince, Director of Special Projects.

Without the cooperation of EIS-related agencies and organizations this book would not be possible. Special thanks go to the following:

Alaska Department of Environmental Conservation: Janice Jennings
American Institute of Certified Planners: Rosemary Jones
Arkansas Department of Pollution Control and Ecology: Jarrell E. Southall,
    Director
California Office of Planning and Research: Ron Bass, Environmental Coordi-
    nator
Center for Natural Areas: Danita C. Mimms
Connecticut Department of Environmental Protection: Stanley J. Pac,
    Commissioner
Delaware Office of Management, Budget and Planning: David S. Hugg, III
Environmental Law Institute: Karen Fishman
Hawaii Office of Environmental Quality Control: Richard L. O'Connell
Kentucky Department for Natural Resources and Environmental Protection,
    Bureau of Environmental Protection: Thomas Grissom, Executive Assistant
    to the Commissioner
Maine Department of Conservation: Ellen Baum
Massachusetts Executive Office of Environmental Affairs: Samuel G. Mygatt
Michigan Pollution Control Agency, Environmental Planning and Review Unit:
    Janet M. Cain
New Jersey Office of Environmental Review: David Applegate, Principal
    Environmental Specialist
New Mexico Environmental Improvement Division: Thomas E. Baca, Director
New York Department of City Planning: Andrew Karn
Ohio Environmental Protection Agency: Beth Whitman
Oklahoma Department of Wildlife Conservation: Richard Gomez
San Francisco Department of City Planning: Gerald Owyang, Office of Environ-
    mental Review
Seattle Department of Community Development: Larry Schmeiser, Director
U.S. Council on Environmental Quality: Rita Tehan
U.S. Department of Agriculture, Forest Service: George Castillo
U.S. Department of Agriculture, Office of Environmental Quality: Glen
    H. Loomis, Associate Director
U.S. Department of Agriculture, Soil Conservation Service: Thomas N. Shiflet,
    Director of Ecological Sciences
U.S. Department of Commerce, Economic Development Administration: Andrew E.
    Kauders, Special Assistant/Environment
U.S. Department of Commerce, National Oceanic and Atmospheric Administration:
    Richard B. Mieremet
U.S. Department of Energy, NEPA Affairs Division: Raymond P. Berube
U.S. Department of Health and Human Services: Charles Custard, Director,
    Office of Environmental Affairs
U.S. Department of the Army, Office of the Chief of Engineers, Environmental
    Programs: George F. Boone, LTC, Assistant Director of Civil Works
U.S. Department of the Interior, Fish and Wildlife Service: A. Gordon Brown,
    Information Officer
U.S. Department of the Interior, Geological Survey: H. William Menard,
    Director
U.S. Department of the Interior, Office of Environmental Affairs, Water
    and Power Resources Service: Al R. Jonez, Director
U.S. Department of the Interior, Office of Environmental Affairs, Water
    and Power Resources Service, Engineering and Research Center: John C.
    Peters

U.S. Department of Transportation, Federal Highway Administration: Leon
H. Larson, Chief, Environmental Programs Division

U.S. Department of Transportation, Office of the Secretary: Martin Convisser,
Director of Environment and Safety

U.S. Environmental Protection Agency, Office of Research and Development:
Patricia L. Cox, Administrative Management Staff

U.S. Environmental Protection Agency, Office of the Administrator: William
N. Hedeman, Jr.

U.S. Environmental Protection Agency, Region I: Wallace E. Stickney, Director,
Environmental and Economic Impact Office

U.S. Environmental Protection Agency, Region II: James R. Marshall, Director
of the Office of External Programs

U.S. Environmental Protection Agency, Region III: Steven A. Torok, Chief,
Environmental Impact Branch

U.S. Environmental Protection Agency, Region IV: John E. Hagan III, Chief,
EIS Branch

U.S. Environmental Protection Agency, Region V: Arlene Kaganove, Environmental Protection Specialist

U.S. Environmental Protection Agency, Region VI: Clinton B. Spotts, Regional
EIS Coordinator

U.S. Environmental Protection Agency, Region X: Roger K. Mochnick, Acting
Chief, Environmental Evaluation Branch

U.S. Federal Energy Regulatory Commission, Environmental Analysis Branch:
Quentin A. Edson, Chief

U.S. Great Lakes Basin Commission: Charles A. Job

U.S. International Boundry and Water Commission, United States and Mexico:
George Baumli, Principal Engineer, Investigations and Planning Division

U.S. National Science Foundation, Office of the Assistant Director for
Astronomical, Atmospheric, Earth, and Ocean Sciences: Adair F. Montgomery,
Chairman, Committee on Environmental Matters

U.S. New England River Basins Commission: Chairman

U.S. Nuclear Regulatory Commission, Division of Engineering, Office of
Nuclear Reactor Regulation: Daniel R. Muller, Assistant Director

U.S. Postal Service, Real Estate and Buildings Department: Robert H. Coven

U.S. Tennessee Valley Authority: Mohamed T. El-Ashry, Director of Environmental Quality

Washington Department of Ecology, Comprehensive Management Division:
Dennis Lundblad, Supervisor

Wisconsin Department of Natural Resources, Bureau of Environmental Impact:
Howard S. Druckenmiller, Director

CONTENTS

EXPLANATORY NOTES

DIRECTORY ORGANIZATION

    This directory was designed to provide the user with two methods of
locating entries: thematically and alphabetically. The thematic arrangement
is divided into three broad classes: (A) General; (B) Physical; and (C) Cultural. These classes are further divided into fourteen EIS themes as follows:

    A. <u>General</u>
            1. General
            2. Legal

    B. <u>Physical</u>
            3. Air
            4. Earth
            5. Noise
            6. Plant/Animal
            7. Water

    C. <u>Cultural</u>
            8. Archaeology/History
            9. Energy/Utilities
           10. Health
           11. Housing
           12. Population
           13. Recreation
           14. Transportation

    The thematic arrangement has been chosen because it provides the user
with: (1) an opportunity to search for agencies and organizations with specific EIS themes; (2) an additional opportunity to search using an alphabet-
ical index; and (3) a more accurate presentation of the theme-by-theme ap-
proach used in EIS analysis and review.

    The second method of locating entries is alphabetical which lists all
entries found in the thematic arrangement. Alphabetization throughout this

book is letter-by-letter. For example, Department of Commerce is followed by Department of the Army.

ENTRY ORGANIZATION

The following presents the organization of individual entries found in this directory.

(1)   Agency/Organization Name
(2)   Street Number/P.O. Box
(3)   City, State/Zip Code
(4)   (Area Code) Telephone Number
(5)   EIS Contact

Key to Entry Organization

(1) Agency/Organization Name. In order to distinguish between governmental agencies and non-governmental organizations, the following rule has been followed: governmental agencies present only the first word of its name on the top line of the entry, while non-governmental organizations list the entire name on the top line of the entry. For example:

Governmental Agency: California
                        Department of Conservation
                        1416 Ninth Street
                        Sacramento, CA 95814
                          (916) 445-8733
                          Director

Non-Governmental   California Tomorrow
Organization: 9681 Market Street
                        San Francisco, CA 94105
                          (415) 391-7544
                          President

(2) Street Number/P.O. Box. This includes all data identifying the location of the entry including building names and in-house addressing codes.

(3) City, State/Zip Code. States follow the U.S. Postal two-letter state and territory abbreviations.

(4) (Area Code) Telephone Number. In some cases, where the information is unavailable the general telephone number is listed rather than the specific telephone number of the EIS contact.

(5) EIS Contact. This directory lists one contact person per entry. Attempts have been made to list the most appropriate person.

DEFINING THE EIS NATIONAL NETWORK

For the purposes of this directory, the following agencies and organizations have been included in the EIS national network:

(1) All federal agencies involved in the EIS process.

(2) National organizations with EIS-related interests.

(3) Major interstate regional planning agencies and organizations.

(4) A brief listing of all 50 states.

(5) A detailed listing of states with comprehensive statutory requirements.

(6) All counties within states with comprehensive statutory requirements.

(7) All major cities within states with comprehensive statutory requirements.

ENVIRONMENTAL IMPACT STATEMENT INSTITUTE

The EIS Institute is a research and literary organization. The Institute is dedicated to: (1) the study of the EIS process; (2) the coordination of information between writers, reviewers, and citizens; (3) the publication of research to the EIS community; and (4) the standardization of EIS terminology and content. Mailing address: 1817 Seashore Drive, Tacoma, WA, 98465.

COMMENT FORM

The EIS Institute welcomes your comments and suggestions concerning this directory. Please fill out the following form and mail to: EIS Institute, 1817 Seashore Drive, Tacoma, WA, 98465.

_____

Entry   Page Number

_____

Comment

_____

Your Name

_____

Your Agency/Organization

_____

Agency/Organization Address

_____

Agency/Organization Telephone Number

_____

GENERAL DIRECTORIES

1. GENERAL EIS-RELATED AGENCIES AND ORGANIZATIONS

Accomack County
Planning Department
County Courthouse
Accomac, VA 23301
  (804) 787-4289
  Director

Accomac-Northampton Planning
  District Commission
Old County Office Building
P.O. Box 316
Accomac, VA 23301
  (804) 787-2936
  Director

Adams County
Planning Department
County Courthouse
Decatur, IN 46733
  (219) 724-4312
  Director

Adams County
Planning Department
165 First Avenue
P.O. Box 334
Othello, WA 99344
  (509) 488-9441
  Director: Randy Martin

Adams County
Planning Department
County Courthouse
Friendship, WI 53934
  (608) 339-7811

  Director

Aitkin County
Planning Department
County Courthouse
Aitkin, MN 56431
  (218) 927-2102
  Director

Alabama Association of County Commissions
100 North Jackson Street
Montgomery, AL 36104
  (205) 263-7594
  Executive Director: O.H. Sharpless

Alabama Association of Soil and Water
  Conservation District Supervisors
P.O. Box 10411
Birmingham, AL 35202
  (205) 263-4007
  Director

Alabama Conservancy
1816 East 28th Avenue, South
Birmingham, AL 35209
  (205) 871-0389
  President: Pat Roys

Alabama Environmental Quality
  Association
3815 Interstate Court
Suite 202
Montgomery, AL 36109
  (205) 277-7050

Chairman: John Bloomer

Alabama League of Municipalities
P.O. Box 1270
Montgomery, AL 36102
    (205) 262-2566
    Executive Director: John Watkins

Alamance Council of Governments
124 West Elm Street
Graham, NC 27253
    (919) 228-1312
    Director

Alamance County
Planning Department
124 West Elm Street
Graham, NC 27253
    (919) 228-1312
    Director

Alameda County
Planning Department
1221 Oak Street
Room 555
Oakland, CA 94612
    (415) 874-0500
    Director

Alaska Center for the Environment
913 West Sixth Avenue
Anchorage, AK 99501
    (907) 274-3621
    President: James Brennan

Alaska Conservation Society
Box 80192
College Branch
Fairbanks, AK 99708
    (907) 452-2240
    Director

Alaska
Department of Environmental Conser-
    vation
Pouch O
Juneau, AK 99811
    (907) 465-2600
    Commissioner: Ernst Mueller

Alaska

Department of Environmental Con-
    servation
Laboratory and Monitoring Section
Pouch OL
Juneau, AK 99811
    (907) 463-2165
    Chief: Tom Trible

Alaska
Department of Environmental Con-
    servation
Northern Region Office
P.O. Box 1601
Fairbanks, AK 99707
    (907) 452-1714
    Supervisor: Douglas Lowery

Alaska
Department of Environmental Con-
    servation
Permit Coordination Section
Pouch O
Juneau, AK 99811
    (907) 465-2670
    Chief: Woody Angst

Alaska
Department of Environmental Con-
    servation
Southcentral Regional Office
338 Denali
Room 1206
Anchorage, AK 99501
    (907) 274-5527
    Supervisor: Kyle Cherry

Alaska
Department of Environmental Con-
    servation
Southeast Regional Office
Box 2420
Juneau, AK 99803
    (907) 789-3151
    Regional Supervisor: Randy Bayliss

Alaska
Department of Environmental Con-
    servation
Valdez Field Office
Drawer 1709
Valdez, AK 99686

(907) 835-4698
Doug Lockwood

Alaska Municipal League
Municipal Building
204 North Franklin
Juneau, AK 99801
  (907) 586-6526
  Director

Albany City
Planning Department
City Hall
Albany, NY 12207
  (518) 472-8111
  Director

Albany County
Planning Department
County Courthouse
Room 201
Albany, NY 12207
  (518) 445-7711
  Director

Albemarle County
Planning Department
County Courthouse
Charlottesville, VA 22901
  (804) 296-5841
  Director

Albemarle Regional Planning and
. Development Commission
102 East Queen Street
P.O. Box 589
Edenton, NC 27932
  (919) 482-8444
  Director

Alexander County
Planning Department
County Courthouse
Taylorsville, NC 28681
  (704) 632-9332
  Director

Alexandria City
Planning Department
City Hall
Alexandria, VA 22314
  (703) 750-6000

Director

Alhambra City
Department of Housing and Community
  Development
111 South First Street
Box 351
Alhambra, CA 91802
  (213) 570-5034
  Director: Norman Yoshihara

Allegany County
Planning Department
P.O. Box 1439
Cumberland, MD 21502
  (301) 777-5922
  Director

Allegany County
Planning Department
County Courthouse
Belmont, NY 14813
  (716) 268-7612
  Director

Alleghany County
Planning Department
County Courthouse
Sparta, NC 28675
  (919) 372-8949
  Director

Alleghany County
Planning Department
County Courthouse
Covington, VA 24426
  (703) 962-4918
  Director

Allen County
Planning Department
200 City-County Building
Fort Wayne, IN 46802
  (219) 423-7211
  Director

Alpine County
Planning Department
County Courthouse
Markleeville, CA 96120
  (916) 694-2281
  Director

Amador County
Planning Department
County Courthouse
Amador City, CA   95601
　　(209) 223-0840
　　Director

Amelia County
Planning Department
P.O. Box A
Amelia Court House, VA   23002
　　(804) 561-3039
　　Director

American Conservation Association
30 Rockefeller Plaza
Room 5425
New York, NY   10020
　　(212) 247-8141
　　President: Laurance Rockefeller

American Consulting Engineers Council
Office of Governmental Affairs
1155-15th Street, N.W.
Suite 713
Washington, DC   20005
　　(202) 296-1780
　　Executive Director: Larry Spiller

American Farm Bureau Federation
225 Touhy Avenue
Park Ridge, IL   60068
　　(312) 399-5700
　　President: Allan Grant

American Institute of Architects
1755 New York Avenue, N.W.
Washington, DC   20006
　　(202) 785-7351
　　Director of Research

American Institute of Biological
　Sciences
1401 Wilson Boulevard
Arlington, VA   22209
　　(703) 527-6776
　　President: Dr. Paul Siegel

American Planning Association
1776 Massachusetts Avenue, N.W.
Washington, DC   20036
　　(202) 872-0611

Executive Director

American Samoa
Office of the Govenor
Environmental Quality Commission
Pago Pago, American Samoa   96920
　　Overseas Operator   633-4116
　　Chairperson: Pati Faiai

American Society of Civil Engineers
345 East 47th Street
New York, NY   10017
　　(212) 644-7491
　　Executive Director: Dr. Eugene
　　Zwoyer

Amherst County
Planning Department
County Courthouse
Amherst, VA   24521
　　(804) 946-7206
　　Director

Anaheim City
Planning Department
Civic Center
P.O. Box 3222
Anaheim, CA   92803
　　(714) 999-5100
　　EIS Contact: Robert Kelly

Anderson City
Planning Department
P.O. Box 2100
Anderson, IN   46011
　　(317) 644-8821
　　Director

Anne Arundel County
Planning Department
Arundel Center
Annapolis, MD   21404
　　(301) 224-1397
　　Director

Anoka County
Planning Department
County Courthouse
Anoka, MN   55303
　　(612) 421-4760
　　Director

Anson County
Planning Department
County Courthouse
Wadesboro, NC 28170
  (704) 694-2796
  Director

Appleton City
Planning Department
City Hall
Appleton, WI 54911
  (414) 733-7329
  Director

Appomattox County
Planning Department
County Courthouse
Appomattox, VA 24522
  (804) 352-5540
  Director

Argonne National Laboratory
Biomedical and Environmental Research Program
9700 South Cass Avenue
Argonne, IL 60439
  (312) 972-3804
  Director

Arizona Association of Conservation
  Districts
Route 1
Box 4
Casa Grande, AZ 85222
  (602) 836-2022
  President: James Henness

Arizona Association of Counties
Room 204
1820 West Washington
Phoenix, AZ 85007
  (602) 252-6563
  Executive Director: Richard Casey

Arizona Conservation Council
Box 11312
Phoenix, AZ 85061
  (602) 262-3716
  Chairman: Howard Gillmore

Arizona

Division of Environmental Health
  Services
1740 West Adams Street
Phoenix, AZ 85007
  (602) 271-4655
  Assistant Director: Bruce Scott

Arizona League of Cities and Towns
1820 West Washington
Phoenix, AZ 85007
  (602) 258-5786
  Executive Director: John DeBolske

Arlington City
Planning Department
City Hall
Arlington, VA 22210
  (703) 558-0200
  Director

Arlington County
Planning Department
County Courthouse
Arlington, VA 22201
  (703) 558-0200
  Director

Arlington Town
Planning Department
Town Hall
Arlington, MA 02174
  (617) 643-6700
  Director

Arkansas Association of Counties
118 National Old Line Building
Little Rock, AR 72201
  (501) 372-7550
  Executive Director: Courtney
    Langston

Arkansas Association of Conservation
  Districts
Bradley, AR 71826
  (501) 894-3472
  President: Andrew Whisenhunt

Arkansas
Department of Pollution Control and
  Ecology
8001 National Drive
P.O. Box 9583

Little Rock, AR 72219
  (501) 371-1701
  Director: Jarrell Southall

Arkansas Municipal League
P.O. Box 38
North Little Rock, AR 72115
  (501) 758-1610
  Executive Director: Don Zimmerman

Arrowhead Regional Development
  Commission
200 Arrowhead Place
Duluth, MN 55802
  (218) 722-5545
  Director

Ashe County
Planning Department
County Courthouse
Jefferson, NC 28640
  (919) 246-8841
  Director

Asheville City
Planning Department
City Hall
Asheville, NC 28807
  (704) 255-5406
  Director

Ashland County
Planning Department
County Courthouse
Ashland, WI 54806
  (715) 682-2533
  Director

Asotin County
Planning Department
County Courthouse
Asotin, WA 99402
  (509) 243-4164
  Director

Association for Conservation Infor-
  mation
Department of Fish, Game and Parks
State Office Building
Pierre, SD 57501
  (605) 224-3485
  President: Chuck Post

Association of Bay Area Governments
Hotel Claremont
Berkeley, CA 94705
  (415) 841-9730
  Program Manager: Michael Visconti

Association of Conservation Engineers
Missouri Department of Conservation
P.O. Box 180
Jefferson City, MO 65101
  (314) 266-2387
  President: Ronald Hansen

Association of Monterey Bay Area
  Governments
1011 Cass Street
P.O. Box 190
Monterey, CA 93940
  (408) 373-8477
  Director

Association of New Jersey Environmen-
  tal Commissions
Box 157
Mendham, NJ 07945
  (201) 539-7547
  President: William Metterhouse

Association of Texas Soil and Water
  Conservation Districts
Box 95
Cherokee, TX 76832
  (915) 622-4227
  President: Kenneth Kuykendall

Atlantic Center for the Environment
951 Highland Street
Ipswich, MA 01938
  (617) 468-4423
  Director: Lawrence Morris

Atlantic County Citizens Council on
  the Environment
9100 Amherst Avenue
Margate, NJ 08402
  (609) 822-3239
  President: Harold Abrams

Augusta County
Planning Department
P.O. Box 488
Staunton, VA 24401

(703) 885-8931
Director

Aurora County
Planning Department
County Courthouse
Plankinton, SD   57368
    (605) 942-9901
    Director

Austin-Mower County Areawide Planning
    Organization
Mower County Courthouse
Austin, MN   55912
    (507) 433-2164
    Director

Avery County
Planning Department
County Courthouse
Newland, NC   28657
    (704) 733-9366
    Director

Bakersfield City
Planning Department
City Hall
Bakersfield, CA   93301
    (805) 861-2767
    Director

Baltimore City
Planning Department
City Hall
Baltimore, MD   21202
    (301) 396-3100
    Director

Baltimore County
Office of Planning and Zoning
Environment Planning Section
Towson, MD   21204
    (301) 494-3211
    EIS Contact: Paul Solomon

Baltimore Environmental Center
Office of the Director
333 East 25th Street
Baltimore, MD   21218
    (301) 366-2070
    Jan Walker

Barnstable County
Planning Department
County Courthouse
Barnstable, MA   02630
    (617) 362-2511
    Director

Barron County
Planning Department
County Courthouse
Barron, WI   54812
    (715) 537-3212
    Director

Bartholomew County
Planning Department
County Courthouse
Columbus, IN   47201
    (812) 372-8818
    Director

Bath County
Planning Department
County Courthouse
Warm Springs, VA   24484
    (703) 839-2361
    Director

Battelle Memorial Institute
Pacific Northwest Laboratories
Ecosystems Department
P.O. Box 999
Richland, WA   99352
    (509) 946-7590
    Dr. Burton Vaughan

Bayfield County
Planning Department
County Courthouse
Washburn, WI   54891
    (715) 373-5508
    Director

Bay-Lake Regional Planning Commission
S.E. Building
University of Wisconsin
Green Bay, WI   54302
    (414) 465-2135
    Director

Beadle County

Planning Department
County Courthouse
Huron, SD 57350
  (605) 352-8436
  Director

Bear Paw Development Corporation
  of Northern Montana
P.O. Box 1549
Havre, MT 59501
  (406) 265-9226
  Director

Beaufort County
Planning Board
P.O. Box 1027
Washington, NC 27889
  (919) 946-0629
  EIS Contact: John Prevette

Beaverhead County
Planning Department
County Courthouse
Dillon, MT 59725
  (406) 683-2642
  Director

Becker County
Planning Department
County Courthouse
Detroit Lakes, MN 56501
  (218) 847-6552
  Director

Bedford County
Planning Department
P.O. Box 234
Bedford, VA 24523
  (703) 586-8813
  Director

Bellevue City
Planning Department
P.O. Box 1768
Bellevue, WA 98009
  (206) 455-6800
    Environmental Coordinator: Diane
      White

Bellflower City
Planning Department
City Hall

Bellflower, CA 90706
  (213) 866-9003
  Director

Beltrami County
Planning Department
County Courthouse
Bemidji, MN 56601
  (218) 751-7300
  EIS Contact: William Patnaude

Bennett County
Planning Department
County Courthouse
Martin, SD 57551
  (605) 685-6591
  Director

Benton County
County Coordinator's Office
County Courthouse
Foley, MN 56329
  (612) 968-7013
  EIS Contact: Al Barthelemy

Benton County
Planning Department
County Courthouse
Fowler, IN 47944
  (317) 884-0760
  Director

Benton County
Planning Department
County Courthouse
Prosser, WA 99350
  (509) 786-2262
  Director: Terry Marden

Benton-Franklin Governmental Conference
1935 Terminal Drive
P.O. Box 217
Richland, WA 99352
  (509) 943-9185
  EIS Contact: Gary Karnofski

Berkeley City
Planning Department
2180 Milvia Street
Berkeley, CA 94704
  (415) 644-6534

Senior Planner: Peter Brady

Berkshire County
Planning Department
County Courthouse
Pittsfield, MA 01201
  (413) 448-8424
  Director

Berkshire County Regional Planning
  Commission
10 Fenn Street
Pittsfield, MA 01201
  (413) 442-1521
  Director

Bertie County
Planning Department
P.O. Box 530
Windsor, NC 27983
  (919) 794-2139
  Director

Big Horn County
Planning Department
County Courthouse
Hardin, MT 59034
  (406) 665-1504
  Director

Big Stone County
Planning Department
County Courthouse
Ortonville, MN 56278
  (612) 839-2105
  Director

Billings City
Planning Department
City Hall
Billings, MT 59103
  (406) 248-7511
  Director

Billings-Yellowstone City-County
  Planning Board
Courthouse
Billings, MT 59101
  (406) 252-5181
  Director

Binghamton City

Planning Department
City Hall
Government Plaza
Binghamton, NY 13901
  (607) 772-7028
  Director: Richard Marko

Blackford County
Planning Department
County Courthouse
Hartford City, IN 47348
  (317) 348-1620
  Director

Black River-St. Lawrence Regional
  Planning Board
Payson Hall
St. Lawrence University
Canton, NY 13617
  (315) 379-5355
  Director

Bladen County
Planning Department
County Courthouse
P.O. Box 1635
Elizabethtown, NC 28337
  (919) 862-2011
  EIS Contact: James Perry

Blaine County
Planning Department
County Courthouse
Chinook, MT 59523
  (406) 357-3240
  Director

Bland County
Planning Department
P.O. Box 276
Bland, VA 24315
  (703) 688-4361
  EIS Contact: James Lucas

Bloomington City
Planning Department
City Hall
Bloomington, MN 55431
  (612) 881-5811
  Director

Blue Earth County

Planning Department
County Courthouse
Mankato, MN 56001
  (507) 625-3031
  Director

Bon Homme County
Planning Department
County Courthouse
Tyndall, SD 57066
  (605) 589-3391
  Director

Boone County
Area Plan Commission
County Courthouse
Lebanon, IN 46052
  (317) 482-3821
  Executive Director: Jerry March

Boston City
Redevelopment Authority
1 City Hall Square
Boston, MA 02108
  (617) 725-4000
  EIS Contact: Richard Mertens

Botetourt County
Planning Department
County Courthouse
Fincastle, VA 24090
  (703) 473-8220
  Director

Bozeman City-County Planning Board
411 East Main
P.O. Box 640
Bozeman, MT 59715
  (406) 586-3321
  Director: Paul Bolton

Bridgeport City
Office of Humane Affairs
45 Lyon Terrace
Bridgeport, CT 06604
  (203) 576-7081
  Environmental Review Coordinator:
    Edward Marshall

Bristol City
Planning Department
City Hall

Bristol, CT 06010
  (203) 583-1811
  Director

Bristol County
Planning Department
County Courthouse
Taunton, MA 02780
  (617) 824-9681
  Director

Bristol Tennessee-Virginia Planning
  Commission
Dickey Building
1009 West State Street
Bristol, VA 24201
  (703) 669-0105
  Director

Broadway County
Planning Department
County Courthouse
Townsend, MT 59644
  (406) 266-3443
  Director

Brockton City
Planning Department
City Hall
Brockton, MA 02401
  (617) 580-1100
  Director

Bronx County
Planning Department
County Courthouse
Bronx, NY 10451
  (212) 293-8000
  Director

Brookings County
Planning Department
County Courthouse
Brookings, SD 57006
  (605) 692-6284
  Director

Brookline Town
Planning Department
Town Hall
Brookline, MA 02146
  (617) 232-9000

Director

Broome County
Environmental Management Council
P.O. Box 1766
Binghamton, NY 13902
  (607) 772-2114
  Environmental Resource Coordinator:
    John Kowalchyk

Brown County
Planning Commission
100 North Jefferson Street
Green Bay, WI 54301
  (414) 497-3633
  Senior Planner: Pat Vaile

Brown County
Planning Department
County Courthouse
Nashville, IN 47448
  (812) 988-2788
  Director

Brown County
Planning Department
County Courthouse
New Ulm, MN 56073
  (507) 354-2215
  Director

Brown County
Planning Department
County Courthouse
Aberdeen, SD 57401
  (605) 622-2266
  Director

Brule County
Planning Department
County Courthouse
Chamberlain, SD 57325
  (605) 734-5443
  Director

Brunswick County
Planning Department
County Courthouse
Southport, NC 28461
  (919) 457-6422
  Director

Brunswick County
Planning Department
P.O. Box 13
Lawrenceville, VA 23868
  (804) 848-3107
  Director

Buchanan County
Planning Department
County Courthouse
Grundy, VA 24614
  (703) 935-2745
  Director

Buckingham County
Planning Department
County Courthouse
Buckingham, VA 23921
  (804) 969-4371
  Director

Buena Park City
Planning Department
City Hall
Buena Park, CA 90620
  (714) 521-9900
  Director

Buffalo City
Planning Department
City Hall
Buffalo, NY 14202
  (716) 856-4200
  Director

Buffalo County
Planning Department
County Courthouse
Grannvalley, SD 57341
  (605) 293-3217
  Director

Buffalo County
Planning Department
County Courthouse
Alma, WI 54610
  (608) 685-4940
  Director

Buncombe County
Planning Department

P.O. Box 7435
Asheville, NC   28807
   (704) 255-5536
   Director

Burbank City
Planning Department
275 East Olive Avenue
P.O. Box 6459
Burbank, CA   91510
   (213) 847-9586
   Principal Planner: Gary Yamada

Burke County
Planning Department
County Courthouse
Morganton, NC   28655
   (704) 437-5721
   Director

Burnett County
Planning Department
County Courthouse
Grantsburg, WI   54840
   (715) 463-5344
   Director

Butte County Association of Governments
7 County Center Drive
Oroville, CA   95965
   (916) 534-4784
   Governmental Planning Coordinator:
     Bill Turpin

Butte City-County Planning Board
Courthouse
Butte, MT   59701
   (406) 723-4714
   Planner: Peggy Delaney

Butte County
Planning Department
P.O. Administration Building
Oroville, CA   95965
   (916) 534-4224
   Director

Butte County
Planning Department
County Courthouse
Belle Fourche, SD   57717

   (605) 892-2516
   Director

Cabarrus County
Planning Department
County Courthouse
Concord, NC   28025
   (704) 786-4137
   Director

Calaveras County
Planning Department
County Courthouse
San Andreas, CA   95249
   (209) 754-4252
   Director

Caldwell County
Planning Department
P.O. Box 1078
Lenoir, NC   28645
   (704) 758-8451
   Director

California Association of Resource
   Conservation Districts
1107 Ninth Street
Room 214
Sacramento, CA   95814
   (916) 447-2535
   President: Lorin Trubschenck

California
Coastal Commission
631 Howard Street
Fourth Floor
San Francisco, CA   94102
   (415) 543-8555
   William Travis

California County Supervisors Association
11th and L Building
Sacramento, CA   95814
   (916) 441-4011
   Executive Director: Richard Watson

California
Department of Conservation
1416 Ninth Street
Room 1354
Sacramento, CA   95814

(916) 445-8733
Janice Moore

California
Department of General Services
1015 L Street
Sacramento, CA 95814
  (916) 445-0780
  Esther Maser

California
Govenor's Office
Office of Planning and Research
1400 Tenth Street
Sacramento, CA 95814
  (916) 322-8515
  Environmental Coordinator: Ron Bass

California League of Cities
1400 K Street
Sacramento, CA 95814
  (916) 444-5790
  Executive Director: Don
    Benninghoven

California
Resources Agency
1416 Ninth Street
Sacramento, CA 95814
  (916) 445-7549
  Bob Stuart

California
State Lands Commission
1807-13th Street
Sacramento, CA 95814
  (916) 322-7813
  Ted Fukushima

California Tomorrow
9681 Market Street
San Francisco, CA 94105
  (415) 391-7544
  President: Weyman Lundquist

Calumet County
Planning Department
County Courthouse
Chilton, WI 53014
  (414) 849-2361
  Director

Calvert County
Planning Department
County Courthouse
Prince Frederick, MD 20678
  (301) 535-1600
  Director

Cambridge City
Planning Department
City Hall
Cambridge, MA 02139
  (617) 876-6800
  Director

Camden County
Planning Department
County Courthouse
Camden, NC 27921
  (919) 335-4077
  Director

Campbell County
Planning Department
County Courthouse
Mound City, SD 57646
  (605) 955-3536
  Director

Campbell County
Planning Department
P.O. Box 100
Rustburg, VA 24588
  (804) 332-5161
  Director

Cape Cod Planning and Economic
  Development Commission
First District Courthouse
Barnstable, MA 02630
  (617) 362-2511
  Director

Cape Fear Council of Governments
One North Third Street
P.O. Box 1491
Wilmington, NC 28401
  (919) 763-0191
  Director

Capital District Regional Planning
  Commission

79 North Pearl Street
Albany, NY 12207
  (518) 474-7444
  Director

Capitol Region Council of Governments
97 Elm Street
Hartford, CT 06106
  (203) 522-2217
  Director

Carbon County
Planning Department
P.O. Box 460
Red Lodge, MT 59068
  (406) 446-1220
  Director: Doug Hart

Carlton County
Zoning Office
County Courthouse
Carlton, MN 55718
  (218) 384-4281
  Zoning Officer: Bruce Benson

Caroline County
Planning Commission
P.O. Box 207
Denton, MD 21629
  (301) 479-2230
  EIS Contact: Alan Visintainer

Caroline County
Planning Department
County Courthouse
Bowling Green, VA 22427
  (804) 633-5380
  Director

Carroll County
Planning Department
County Courthouse
Delphi, IN 46923
  (317) 564-4485
  Director

Carroll County
Planning Department
County Courthouse
Westminster, MD 21157
  (301) 848-4500
  Director

Carroll County
Planning Department
County Courthouse
Hillsville, VA 24343
  (703) 728-3331
  Director

Carson City
Planning Department
City Hall
Carson City, CA 90745
  (213) 830-7600
  Director

Carter County
Planning Department
County Courthouse
Ekalaka, MT 59324
  (406) 775-8714
  Director

Carteret County
Planning Department
County Courthouse
Beaufort, NC 28516
  (919) 728-3644
  Director

Carver County
Planning Department
600 East Fourth
Chaska, MN 55318
  (612) 448-3435

Cascade County
Planning Department
County Courthouse
Great Falls, MT 59401
  (406) 761-6700
  Director

Cass County
Planning Department
County Courthouse
Logansport, IN 46947
  (219) 753-2916
  Director

Cass County
Planning Department
County Courthouse
Walker, MN 56484

(218) 547-3300
Zoning Administrator: Carol
  Newstrand

Caswell County
Planning Department
County Courthouse
Yanceyville, NC 27379
  (919) 694-4171
  Director

Catawba County
Planning Department
P.O. Box 389
Newton, NC 28658
  (704) 464-7880
  EIS Contact: Jack Matthews

Catskill Center for Conservation and
  Development
Hobart, NY 13788
  (607) 538-3581
  President: Dr. Sherret Chase

Cattaraugus County
Planning Department
County Courthouse
Little Valley, NY 14755
  (716) 938-9111
  Director

Cayuga County
Planning Department
County Courthouse
Auburn, NY 13021
  (315) 235-1271
  Director

Cecil County
Planning Department
County Courthouse
Elkton, MD 21921
  (301) 398-4100
  Director

Center for International Environmental Information
300 East 42nd Street
New York, NY 10017
  (212) 697-3232
  Executive Director: Whitman Bassow

Center for Natural Areas
1525 New Hampshire Avenue, N.W.
Washington, DC 20036
  (202) 265-0066
  President: William Reed

Center for Urban Environmental
  Studies
1012-14th Street, N.W.
Suite 906
Washington, DC 20005
  (202) 347-6020
  Executive Director: Larry Young

Central Connecticut Regional Planning
  Agency
12 Landry Street
Bristol, CT 06010
  (203) 589-7820
  Director

Centralina Council of Governments
P.O. Box 4168
Charlotte, NC 28204
  (704) 372-2416
  Director

Central Massachusetts Regional Planning Commission
71 Elm Street
Worcester, MA 01609
  (617) 756-7717
  Director

Central Minnesota Regional Development Commission
2700 First Street North
St. Cloud, MN 56301
  (612) 253-7870
  Director

Central Montana District Six Council
Box 282
Roundup, MT 59072
  (406) 323-2549
  Agency Planner: Dennis Balyeat

Central Naugatuck Valley Regional
  Planning Agency
20 East Main Street
Waterbury, CT 06702

(203) 757-0535
Executive Director: Duncan Graham

Central New York Regional Planning
    and Development Board
Midtown Plaza
700 East Water Street
Syracuse, NY 13210
    (315) 422-8276
    Director: Barbara Blanchard-
    Whispell

Central Shenandoah Planning District
    Commission
119 West Frederick Street
P.O. Box 1337
Staunton, VA 24401
    (703) 885-5174
    Regional Planner: Talmage Reeves

Central Sierra Planning Council
520 North Main Street
P.O. Box 816
Altaville, CA 95221
    (209) 736-0108
    Director

Central Virginia Planning District
    Commission
2511 Memorial Avenue
Lynchburg, VA 24501
    (804) 845-3491
    Director

Chamber of Commerce of the United
    States of America
Resources and Environmental Quality
    Division
1615 H Street, N.W.
Washington, DC 20062
    (202) 659-6172
    Manager: John Robinson

Charles City County
Planning Department
County Courthouse
Charles City, VA 23030
    (804) 829-2402
    Director

Charles County
Planning Department

County Courthouse
P.O. Box B
La Plata, MD 20646
    (301) 934-8141
    Director

Charles Mix County
Planning Department
County Courthouse
Lake Andes, SD 57356
    (605) 487-7511
    EIS Contact: Elvern Varilek

Charlotte City
Planning Department
City Hall
Charlotte, NC 28202
    (704) 274-2040
    Director

Charlotte County
Planning Department
County Courthouse
Charlotte Courthouse, VA 23923
    (804) 542-5147
    Director

Chatham County
Planning Department
P.O. Box 54
Pittsboro, NC 27312
    (919) 542-4873
    Director: Keith Megginson

Chautauqua County
Planning Department
County Office Building
Mayville, NY 14757
    (716) 753-4296
    Director: John Luensman

Chelan County
Planning Department
County Courthouse
Wenatchee, WA 98801
    (509) 663-4803
    Director

Chelan County Regional Planning
    Council
Courthouse Annex
411 Washington Street

Wenatchee, WA 98801
  (509) 663-2101
  Director

Chemung County
Planning Department
County Courthouse
Elmira, NY 14901
  (607) 737-2912
  Director

Chenango County
Planning Department
County Courthouse
Norwich, NY 13815
  (607) 335-4500
  Director

Cherokee County
Planning Department
County Courthouse
Murphy, NC 28906
  (704) 837-5527
  Director

Chesapeake City
Planning Department
City Hall
Chesapeake, VA 23320
  (804) 547-6345
  Director

Chesterfield County
Planning Department
County Courthouse
Chesterfield, VA 23832
  (804) 748-1211
  Director

Chicopee City
Planning Department
City Hall
Chicopee, MA 01013
  (413) 594-4711
  Director

Chippewa County
Planning Department
County Courthouse
Montevideo, MN 56265
  (612) 269-7447
  Director

Chippewa County
Planning Department
County Courthouse
Chippewa Falls, WI 54729
  (715) 723-4168
  Director

Chisago County
Zoning and Building Department
County Courthouse
Box 202
Center City, MN 55012
  (612) 257-1300
  EIS Contact: Gerald Peterson

Chouieau County
Planning Department
County Courthouse
Fort Benton, MT 59442
  (406) 622-5024
  Director

Chowan County
Planning Department
County Courthouse
Edenton, NC 27932
  (919) 482-8431
  Director

Chula Vista City
Planning Department
276 Fourth Avenue
Chula Vista, CA 92010
  (714) 575-5101
  Environmental Review Coordinator:
    Douglas Reid

Clallam County
Planning Department
223 East 4th
Port Angeles, WA 98362
  (206) 452-2102
  EIS Contact: Dave Burns

Clallam County Governmental
  Conference
P.O. Box 430
Port Angeles, WA 98362
  (206) 457-4562
  Director

Clark County

Planning Department
County Courthouse
Jeffersonville, IN   47130
  (812) 283-4451
  Director

Clark County
Planning Department
County Courthouse
Clark, SD   57225
  (605) 532-5851
  Director

Clark County
Planning Department
County Courthouse
Vancouver, WA   98660
  (206) 699-2000
  Director

Clark County
Planning Department
County Courthouse
Neillsville, WI   54456
  (715) 743-3301
  Director

Clark County Regional Planning
  Council
1408 Franklin Street
Environmental Planning Services
P.O. Box 5000
Vancouver, WA   98663
  (206) 699-2361
  Director: Rich Hines

Clarke County
Planning Department
County Courthouse
Berryville, VA   22611
  (703) 955-1309
  County Administrator: Bob Lee

Clay County
Planning Department
County Courthouse
Brazil, IN   47834
  (812) 448-8727
  Director

Clay County
Planning Department

County Courthouse
Moorhead, MN   56560
  (218) 233-2781
  Director

Clay County
Planning Department
County Courthouse
Hayesville, NC   28904
  (704) 389-6411
  Director

Clay County
Planning Department
County Courthouse
Vermillon, SD   57609
  (605) 624-3371
  Director

Clearwater County
Planning Department
County Courthouse
Bagley, MN   56621
  (218) 694-6177
  Director

Cleveland County
Planning Department
P.O. Box 1210
Shelby, NC   28150
  (704) 482-8311
  Director

Clinton County
Planning Department
County Courthouse
Frankfort, IN   46041
  (317) 654-8529
  Director

Clinton County
Planning Department
County Courthouse
Plattsburgh, NY   12901
  (518) 561-8800
  Director

Codington County
Planning Department
County Courthouse
Watertown, SD   57201
  (605) 886-4850

Director

Colorado Counties
Suite 301
1500 Grant Street
Denver, CO 80203
　(303) 861-4076
　Executive Director: Harry Bowes

Colorado
Department of Health
4210 East 11th Avenue
Denver, CO 80220
　(303) 320-8333
　Executive Director: Dr. Frank
　　Traylor

Colorado
Department of Health
Environmental Programs
4210 East 11th Avenue
Denver, CO 80220
　(303) 320-8333
　Associate Director: William
　　Auberle

Colorado Municipal League
4800 Wadsworth Boulevard
Suite 204
Wheat Ridge, CO 80033
　(303) 421-8630
　Executive Director: Kenneth Bueche

Colorado Open Space Council
1325 Delaware Street
Denver, CO 80204
　(303) 573-9241
　President: Terry Stewart

Columbia County
Planning Board
71 North Third Street
Hudson, NY 12534
　(518) 828-3375
　Director: Edith Mesick

Columbia County
Planning Department
County Courthouse
Dayton, WA 99328
　(509) 382-4542
　Director

Columbia County
Planning Department
County Courthouse
Portage, WI 53901
　(608) 742-2191
　Director

Columbia Falls City-County Planning
　Board
Box 417
Columbia Falls, MT 59912
　(406) 892-5788
　Director

Columbus County
Planning Department
111 Washington Street
Whiteville, NC 28472
　(919) 642-3860
　Director

Colusa County
Planning Department
County Courthouse
Colusa, CA 95932
　(916) 458-4660
　Director

Commonwealth of Northern Mariannas
Department of Public Health and
　Environmental Services
Division of Environmental Quality
Dr. Torres Hospital
Saipan, CM 96950
　Director: George Chan

Comprehensive Planning Organization
　of the San Diego Region
Security Pacific Plaza
1200 Third Avenue
San Diego, CA 92101
　(714) 233-5211
　Director

Compton City
Planning Department
City Hall
Compton, CA 90224
　(213) 537-8000

Concord City
Planning Department

Concord Civic Center
1950 Parkside Drive
Concord, CA 94519
    (415) 671-3150
    EIS Contact: David Golick

Connecticut Association of Conser-
    vation Commissions
P.O. Box 571
Old Lyne, CT 06371
    (203) 443-8695
    President: Mervin Roberts

Connecticut Association of Soil and
    Water Conservation Districts
Box 27
Westport, CT 06880
    (203) 227-9265
    President: Albert Kelly

Connecticut Conference of Municipal-
    ities
956 Chapel Street
New Haven, CT 06510
    (203) 772-2168
    Executive Director: Joel Cogen

Connecticut
Council on Environmental Quality
Room 141
165 Capitol Avenue
Hartford, CT 06115
    (203) 566-3510
    Executive Director: Mary Dickinson

Connecticut
Department of Environmental Protec-
    tion
State Office Building
165 Capitol Avenue
Hartford, CT 06115
    (203) 566-5599
    Commissioner: Stanley Pac

Connecticut
Department of Environmental Protec-
    tion
Planning and Coastal Management Unit
State Office Building
Room 118
Hartford, CT 06115
    (203) 566-3740

Principal Environmental Analyst:
    Jonathan Clapp

Connecticut River Estuary Regional
    Planning Agency
Westbrook Road
P.O. Box 335
Essex, CT 06426
    (203) 767-0944
    Director

Conservation and Research Foundation
Box 1445
Connecticut College
New London, CT 06320
    (203) 873-8514
    President: Richard Goodwin

Conservation Associates
1500 Mills Tower
220 Bush Street
San Francisco, CA 94104
    President: Dorothy Varian

Conservation Council for Hawaii
Box 2923
Honolulu, HI 96802
    President: Peter Galloway

Conservation Council of North
    Carolina
307 Granville Road
Chapel Hill, NC 27514
    (919) 942-7935
    President: Dave Martin

Conservation Council of Virginia
Westwood Building
Room 205
2317 Westwood Avenue
Richmond, VA 23230
    (804) 272-2126
    President: Robert Hicks, Jr.

Conservation Federation of Missouri
312 East Capitol Avenue
Jefferson City, MO 65101
    (314) 634-2322
    President: E.J. Seidler

Conservation Foundation
1717 Massachusetts Avenue, N.W.

Washington, DC 20036
  (202) 797-4300
  President: William Reilly

Conservation Services
South Great Road
Lincoln, MA 01773
  (617) 259-9500
  Executive Director: Allen Morgan

Conservation Society of Southern
  Vermont
P.O. Box 256
Townshend, VT 05353
  (802) 365-7754
  President: Peter Strong

Conservation Trust of Puerto Rico
P.O. Box 4747
San Juan, PR 00905
  (809) 722-5834
    Executive Director: Francisco
    Blanco

Contra Costa County
Planning Department
651 Pine Street
Martinez, CA 94553
  (415) 372-4080
  Director

Cook County
Planning Department
County Courthouse
Grand Marais, MN 55604
  (218) 387-1230
  Director

Corson County
Planning Department
County Courthouse
McIntosh, SD 57641
  (605) 273-4201
  Director

Cortland County
Planning Department
County Courthouse
Cortland, NY 13045
  (607) 753-9374
    Environmental Planner: Ronald
    Slotkin

Costa Mesa City
Planning Department
P.O. Box 1200
Costa Mesa, CA 92626
  (714) 556-5327
  Associate Planner: Reba Tovw

Cottonwood County
Planning Department
County Courthouse
Windom, MN 56101
  (507) 831-1905
  Director

Council for Planning and Conservation
Box 228
Beverly Hills, CA 90213
  (213) 276-3202
  President: Ellen Harris

Council of Fresno County Governments
2014 Tulare Street
Suite 520
Fresno, CA 93721
  (209) 233-4148
  EIS Contact: Georgiena Stine

Council of State Governments
Environmental Office
P.O. Box 11910
Lexington, KY 40511
  (606) 252-2291
  Director

Covington County
Planning Department
County Courthouse
Covington, VA 24426
  (703) 962-4984
  Director

Cowlitz County
Planning Department
207 Fourth Avenue, North
Kelso, WA 98626
  (206) 577-3052
  EIS Contact: Sarah Deatherage

Cowlitz-Wahkiakum Governmental Con-
  ference
Fifth Avenue Annex Courthouse
Kelso, WA 98628

(206) 577-3041
Director

Craig County
Planning Department
P.O. Box 206
New Castle, VA 24127
  (703) 864-5010
  Director

Crater Planning District Commission
P.O. Box 1808
Petersburg, VA 23803
  (804) 861-1666
  Director

Craven County
Planning Department
Drawer R
New Bern, NC 28560
  (919) 633-1353
  Director: Donald Baumgardner

Crawford County
Planning Department
County Courthouse
English, IN 47118
  (812) 338-2565
  Director

Crawford County
Planning Department
County Courthouse
Prairie Du Chien, WI 53821
  (608) 326-2122
  Director

Crow Wing County
Planning Department
County Courthouse
Brainerd, MN 56401
  (218) 829-3667
  Director

Culpeper County
Planning Department
County Courthouse
Culpeper, VA 22701
  (703) 825-3035
  Director

Cumberland County

Planning Department
P.O. Drawer 1829
Fayetteville, NC 28301
  (919) 483-8131
  Director

Cumberland County
Planning Department
County Courthouse
Cumberland, VA 23040
  (804) 492-4442
  Director

Cumberland Plateau Planning District
  Commission
Box 548
Lebanon, VA 24266
  (703) 889-1778
  Director

Currituck County
Planning Department
P.O. Box 93
Currituck, NC 27929
  (919) 232-2075
  Director

Custer County
Planning Department
County Courthouse
Miles City, MT 59301
  (406) 232-1347
  Director

Custer County
Planning Department
County Courthouse
Custer, SD 57730
  (605) 673-4816
  Director

Dakota County
Planning Services
Government Center
1560 Highway 55
Hastings, MN 55033
  (612) 437-0225
  Director: Jeffrey Connell

Daly City
Department of Community Development
Sullivan Avenue and 90th Street

Daly City, CA 94015
(415) 992-4500
Director

Danbury City
Environmental Impact Commission
City Hall
Danbury, CT 06810
(203) 744-7160
Inspector: Arthur Bohan

Dane County
Planning Department
County Courthouse
Madison, WI 53709
(608) 266-4121
Director

Dane County Regional Planning
  Commission
City-County Building
Madison, WI 53709
(608) 266-4137
Director

Daniels County
Planning Department
County Courthouse
Scobey, MT 59263
(406) 487-5561
Director

Dare County
Planning Department
County Courthouse
Manteo, NC 27954
(919) 473-2950
Director

Davidson County
Planning Department
County Courthouse
Lexington, NC 27292
(704) 246-2549
Director

Davie County
Planning Department
County Courthouse
Mocksville, NC 27028
(704) 634-5513
EIS Contact: Jesse Boyce, Jr.

Daviess County
Planning Department
County Courthouse
Washington, IN 47501
(812) 254-2713
Director

Davison County
Planning Department
County Courthouse
Mitchell, SD 57301
(605) 996-2450
Director

Dawson County
Planning Department
County Courthouse
Glendive, MT 59330
(406) 365-3058
Director

Day County
Planning Department
County Courthouse
Webster, SD 57274
(605) 345-3102
Director

Dearborn County
Planning Department
County Courthouse
Lawrenceburg, IN 47025
(812) 537-2151
Director

Decatur County
Planning Department
County Courthouse
Greensburg, IN 47240
(812) 663-8223
Director

Deer Lodge County
Planning Department
County Courthouse
Anaconda, MT 59711
(406) 563-6541
Director

De Kalb County
Planning Department
County Courthouse

Auburn, IN 46706
  (219) 925-0912
  Director

Delaware Association of Conservation
  Districts
Ellendale, DE 19941
  (302) 422-8698
  President: Elwood Tucker

Delaware Association of Counties
New Castle County Public Building
Wilmington, DE 19801
  (302) 571-7520
  Executive Director: Joseph Toner

Delaware County
Planning Department
County Courthouse
Muncie, IN 47302
  (317) 747-7726
  Director

Delaware County
Planning Department
County Courthouse
Delhi, NY 13753
  (607) 746-2603
  Director

Delaware
Department of Natural Resources and
  Environmental Control
Tatnall Building
P.O. Box 1401
Dover, DE 19901
  (302) 678-4764
  Director: Edward Sienicki

Delaware
Department of Natural Resources and
  and Environmental Control
Division of Environmental Control
Tatnall Building
P.O. Box 1401
Dover, DE 19901
  (302) 678-4765
  Tom Eichler

Delaware
Department of Natural Resources and
  Environmental Control

Division of Soil and Water Conser-
  vation
Tatnall Building
P.O. Box 1401
Dover, DE 19901
  (302) 678-4411
  Director: William Ratledge

Delaware League of Local Governments
P.O. Box 484
Dover, DE 19901
  (302) 678-0991
  Executive Director: Leon de Valinger

Delmarva Advisory Council
One Plaza East
P.O. Box 711
Salisbury, MD 21801
  (301) 742-9271
  Director

Del Norte County
Planning Department
700-5th Street
Cresent City, CA 95531
  (707) 464-2119
  County Planner: Ernest Perry

Desert Protective Council
P.O. Box 4294
Palm Springs, CA 92263
  (714) 469-5179
  President: Harriet Allen

Deuel County
Planning Department
County Courthouse
Clear Lake, SD 57226
  (605) 874-2120
  Director

Dewey County
Planning Department
County Courthouse
Timber Lake, SD 57656
  (605) 865-3566
  County Auditor: Harry Hodgman

Dickenson County
Planning Department
County Courthouse
Clintwood, VA 24228

(703) 926-4549
Director

Dinwiddie County
Planning Department
P.O. Box 400
Dinwiddie, VA 23841
 (804) 469-2611
 Director

District of Columbia
Department of Environmental Services
415-12th Street, N.W.
Washington, DC 20004
 (202) 629-3415
 Director: Herbert Tucker

Dodge County
Planning Department
County Courthouse
Mantorville, MN 55955
 (507) 635-3541
 EIS Contact: Cec Samuelson

Dodge County
Planning Department
County Courthouse
Juneau, WI 53039
 (414) 386-4411
 Director

Door County
Planning Department
County Courthouse
Sturgeon Bay, WI 54235
 (414) 743-5511
 Director: Robert Florence

Dorchester County
Planning Department
County Courthouse
Cambridge, MD 21613
 (301) 228-1700
 Director

Douglas County
Planning Department
County Courthouse
Alexandria, MN 56308
 (612) 763-6076
 Director

Douglas County
Planning Department
County Courthouse
Armour, SD 57313
 (605) 724-2585
 Director

Douglas County
Planning Department
County Courthouse
Waterville, WA 98858
 (509) 745-3001
 Director

Douglas County
Planning Department
County Courthouse
Superior, WI 54880
 (715) 394-0341
 Director

Douglas County Regional Planning
  Commission
110 Third Street, N.E.
East Wenatchee, WA 98801
 (509) 884-7221
 Director

Downey City
Planning Department
City Hall
Downey, CA 90241
 (213) 861-0361
 Director

Dubois County
Planning Department
County Courthouse
Jasper, IN 47546
 (812) 482-5445
 Director

Dukes County
Planning Department
P.O. Box 268
Edgartown, MA 02539
 (617) 627-5535
 Director

Duluth City
Planning Department
City Hall

Duluth, MN 55801
  (218) 723-3340
  Director

Dunn County
Planning Department
800 Wilson Avenue
Menomonie, WI 54751
  (715) 232-2429
  Director

Duplin County
Planning Department
P.O. Box 158
Kenansville, NC 28349
  (919) 296-1686
  Director

Durham City
Planning Department
101 City Hall Plaza
Durham, NC 27701
  (919) 683-4100
  EIS Contact: Annette Liggett

Durham County
Planning Department
County Office Building
Durham, NC 27701
  (919) 688-3360
  Director

Dutchess County
Planning Department
47 Cannon Street
Poughkeepsie, NY 12601
  (607) 485-9890
  EIS Contact: Eric Gillert

East Central Regional Development
  Commission
119 South Lake Street
Mora, MN 55051
  (612) 679-4065
  EIS Contact: Bob Plilford

East Central Wisconsin Regional
  Planning Commission
1919 American Court
Neenah, WI 54946
  (414) 739-6156
  Director

East Hartford Town
Planning Department
Town Hall
East Hartford, CT 06108
  (203) 289-2781
  Director

Eau Claire County
Planning Department
County Courthouse
Eau Claire, WI 54701
  (715) 839-4801
  Director

Ecological Society of America
Department of Botany
Duke University
Durham, NC 27706
  (919) 684-5544
  President: Dr. Dwight Billings

Ecology Action for Rhode Island
286 Thayer Street
Providence, RI 02906
  (401) 274-9429
  President: Barry Schiller

Ecology Center
2701 College Avenue
Berkeley, CA 44705
  (415) 548-2220
  Chairperson: Kris Muller

Ecology Center of Louisiana
Office of the President
P.O. Box 19344
New Orleans, LA 70179
  (504) 482-8760
  President: Ross Vincent

Edgecombe County
Planning Department
County Courthouse
Tarboro, NC 27886
  (919) 823-8121
  Director

Edmunds County
Planning Department
County Courthouse
Ipswich, SD 57451
  (605) 426-3231

Director

El Cajon City
Planning Department
200 East Main Street
El Cajon, CA 92020
(714) 440-1776
EIS Contact: Barbara Ramirez

El Dorado County
Planning Department
360 Fair Lane
Placerville, CA 95667
(916) 626-2371
Principal Planner: Jake Raper

Elkhart County
Planning Department
County Courthouse
Goshen, IN 46526
(219) 533-3610
Director

El Monte City
Planning Department
City Hall
El Monte, CA 91734
(213) 575-2225
Director

Environment
Committee for Environmental Information
560 Trinity Boulevard
St. Louis, MO 63130
(314) 727-3311
Director

Environmental Action
1346 Connecticut Avenue, N.W.
Room 731
Washington, DC 20036
(202) 833-1845
Staff: Peter Harnik

Environmental Action Coalition
516 Fifth Avenue
New York, NY 10010
(212) 929-8481
Executive Director: Seymour
Josephson

Environmental Action Foundation
724 DuPont Circle Building
Washington, DC 20036
(202) 659-9682
President: Liz Tennant

Environmental Centers
950 Trout Brook Drive
West Hartford, CT 06119
(203) 236-2961
Director

Environmental Coalition
59 North Main Street
Box 757
Concord, NH 03301
(603) 224-7575
Chairman: Coburn Wheeler

Environmental Fund
1302-18th Street, N.W.
Washington, DC 20036
(202) 293-2548
Director: Justin Blackwelder

Environmental Industry Council
Office of the President
1825 K Street, N.W.
Washington, DC 20006
(202) 331-7706
John Adams

Environmental Information Center
P.O. Box 1184
Helena, MT 59601
(406) 443-2520
Staff Coordinator: Bob Kiesling

Environmental Information Center of
the Florida Conservation Foundation
935 Orange Avenue
Winter Park, FL 32789
(305) 644-5377
President: Harlan Herbert

Environmental Planning Lobby
196 Morton Avenue
Albany, NY 12202
(518) 462-5526
President: Henry Neale, Jr.

Environmental Policy Center

317 Pennsylvania Avenue, S.E.
Washington, DC   20003
  (202) 547-6500
  President: Carolyn Alderson

Environmental Research Institute
Box 156
Moose, WY 83012
  President: Frank Craighead, Jr.

Environmental Resources
Northeastern Illinois University
Department of Earth Sciences
Bryn Mawr at St. Louis
Chicago, IL 60625
  (312) 583-4050
  Director

Environmental Study Conference
Room 3349
House Office Building
Annex 2
Washington, DC 20515
  (202) 255-2988
  Director

Environment Council of Rhode Island
40 Bowen Street
Providence, RI 02903
  (401) 521-1670
  President: Robert Harpell

Environment Forum
5606 Vernon Place
Bethesda, MD 20034
  (301) 654-5524
  Harold Leich

Environment Information Center
292 Madison Avenue
New York, NY 10017
  (212) 949-9494
  President: James Kollegger

Envirosouth
P.O. Box 17111
Montgomery, AL 36117
  (205) 277-7050
  President: Martha McInnis

Erie and Niagara Counties Regional
  Planning Board

Northtown Plaza
3103 Sheridan Drive
Amherst, NY 14226
  (716) 837-2035
  Director: Leo Nowak, Jr.

Erie County
Department of Environment and
  Planning
95 Franklin Street
Buffalo, NY 14202
  (716) 846-8390
  Commissioner: Joan Loring

Essex County
Planning Department
32 Federal Street
Alem, MA 01970
  (617) 744-2840
  Director

Essex County
Planning Department
County Courthouse
Elizabethtown, NY 12932
  (518) 873-6301
  Director: William Johnston

Essex County
Planning Department
P.O. Box 1079
Tappahannock, VA 22560
  (804) 443-4331
  Director

Evansville City
Planning Department
302 Civic Center Complex
Evansville, IN 47708
  (812) 426-5000
  Director

Everett City
Planning Department
City Hall
Everett, WA 98201
  (206) 259-8755
  Director

Fairbault County
Planning Department
County Courthouse

Blue Earth, MN   56013
   (507) 526-5145
   Director

Fairfax County
Planning Department
4100 Chain Bridge Road
Fairfax, VA   22030
   (703) 691-2321
   Director

Fairfield Town
Planning Department
Town Hall
Fairfield, CT   06430
   (203) 259-8361
   Director

Fallon County
Planning Department
County Courthouse
Baker, MT   59313
   (406) 778-2846
   Director

Fall River City
Planning Department
City Hall
Fall River, MA   02720
   (617) 675-6011
   Director

Fall River County
Planning Department
County Courthouse
Hot Springs, SD   57747
   (605) 745-5131
   Director

Fargo-Moorhead Metropolitan Council
   of Governments
44 Foss Lane
Moorhead, MN   56560
   (218) 233-2704
   Director

Faulk County
Planning Department
County Courthouse
Faulkton, SD   57438
   (605) 598-6223
   Director

Fauquier County
Planning Department
County Courthouse
Warrenton, VA   22186
   (703) 347-9550
   Director

Fayette County
Planning Department
County Courthouse
Connersville, IN   47331
   (317) 825-1813
   Director

Fayetteville City
Planning Department
City Hall
Fayetteville, NC   28302
   (919) 483-4577
   Director

Federation of Rocky Mountain States
2480 West 26th Avenue
Suite 300-B
Denver, CO   80211
   (303) 458-8000
   President

Fergus County
Planning Department
County Courthouse
Lewistown, MT   59457
   (406) 538-5321
   Director

Ferry County
Planning Department
P.O. Box 305
Republic, WA   99166
   (509) 775-3705
   Director: Carl Putnam

Fillmore County
Planning Department
County Courthouse
Preston, MN   55965
   (507) 765-4701
   Director

Five Valleys Council of Governments
Missoula County Courthouse
Missoula, MT   59801

(406) 728-0820
Director

Flathead County
Planning Department
County Courthouse
Kalispell, MT 59901
  (406) 755-5300
  Director

Flathead County Areawide Planning
  Organization
Flathead County Health Service Center
723 Fifth Avenue East
Room 414
Kalispell, MT 59901
  (406) 755-5300
  Senior Planner: Gary Hill

Florence County
Planning Department
County Courthouse
Florence, WI 54121
  (715) 528-3201
  Director

Florida Association of Soil and Water
  Conservation District Supervisors
P.O. Box 301
LaBelle, FL 33935
  (813) 675-2461
  President: J.R. Paul, Jr.

Florida Conservation Council
1111 North Magnolia
Apartment M101
Tallahassee, FL 32301
  (904) 488-1686
  Chairman: T.N. Anderson

Florida
Department of Environmental Regulations
2600 Blairstone Road
Tallahassee, FL 32301
  (904) 488-4807
  Chairman: W.D. Frederick, Jr.

Florida League of Cities
P.O. Box 1757
Tallahassee, FL 32302
  (904) 222-9684

Executive Director: Raymond Sittig

Florida State Association of County
  Commissioners
P.O. Box 549
Tallahassee, FL 32302
  (904) 224-3148
  Executive Director: John Thomas

Floyd County
Plan Commission
Suite 216
City-County Building
New Albany, IN 47150
  (812) 945-1315
  Executive Director: Patrick Houghlin

Floyd County
Planning Department
P.O. Box 53
Floyd, VA 24091
  (703) 745-2610
  Director

Fluvanna County
Board of Supervisors
P.O. Box 137
Palmyra, VA 22963
  (804) 589-3138
  County Administrator: Gregory
    Wolfrey

Fond Du Lac County
Planning Department
County Courthouse
Fond Du Lac, WI 54935
  (414) 921-5600
  Director

Foresta Institute for Ocean and
  Mountain Studies
6205 Franktown Road
Carson City, NV 89701
  (702) 882-6361
  Nancy Gaffney

Forest County
Planning Department
County Courthouse
Crandon, WI 54520
  (715) 478-2422
  Director

Forsyth County
Planning Department
Hall of Justice
Winston-Salem, NC 27101
  (919) 761-2250
  Director

Fort Wayne City
Planning Department
City-County Building
One Main Street
Fort Wayne, IN 46802
  (219) 423-7211
  Senior Planner: David Baker

Fountain County
Planning Department
County Courthouse
Covington, IN 47932
  (317) 793-2192
  Director

Framingham Town
Planning Department
Town Hall
Framingham, MA 01701
  (617) 879-8570
  EIS Contact: Larry Dennison

Franklin County
Planning Department
County Courthouse
Brookville, IN 47012
  (317) 647-4631
  Director

Franklin County
Planning Department
County Courthouse
Greenfield, MA 01301
  (413) 774-4015
  Director

Franklin County
Planning Department
County Courthouse
Malone, NY 12953
  (518) 483-2761
  Director

Franklin County
Planning Department

113 Market Street
Louisburg, NC 27549
  (919) 496-5104
  Director

Franklin County
Planning Department
302 Virgil H. Goode Building
Rocky Mount, VA 24151
  (703) 483-1315
  County Administrator: Billy Beckett

Franklin County
Planning Department
1016 North Fourth Street
Pasco, WA 99301
  (509) 545-3535
  Director: Robert Boothe

Frederick County
Planning Department
Winchester Hall
Frederick, MD 21701
  (301) 694-1149
  EIS Contact: Mark Friis

Frederick County Council of Govern-
  ments
Winchester Hall
Frederick, MD 21701
  (301) 663-8300
  Director

Frederick County
Planning Department
9 Court Square
Winchester, VA 22601
  (703) 667-2365
  Director

Freeborn County
Planning and Zoning Office
County Courthouse
Albert Lea, MN 56007
  (507) 373-0628
  EIS Contact: Truman Thrond

Fremont City
Planning Department
City Government Building
Fremont, CA 94538
  (415) 791-4165

Senior Planner: Bob Carlson

Fresno City
Planning and Inspection Department
City Hall
2326 Fresno Street
Fresno, CA 93721
  (209) 488-1371
  EIS Contact: Robert Dyer

Fresno County
Planning Department
County Courthouse
Fresno, CA 93721
  (209) 488-3033
  Director

Friends of the Earth
124 Spear Street
San Francisco, CA 94105
  (415) 495-4770
  President: David Brower

Friends of the Earth
Alaska Representative
1895 Pioneer Way
Fairbanks, AK 99701
  (907) 479-6077
  Representative: Jim Kowalsky

Friends of the Earth
Colorado Plateau Region
P.O. Box 820
Moab, UT 84532
  Representative: Gordon Anderson

Friends of the Earth
Mid-Atlantic Region
72 Jane Street
New York, NY 10014
  (212) 675-5911
  Representative: Lorna Salzman

Friends of the Earth
Midwest Region
P.O. Box 31
Columbia, MO 62501
  (314) 449-4721
  Representative: Don Pierce

Friends of the Earth
New England Region

3 Joy Street
Boston, MA 02108
  (617) 742-6329
  Representative: Ann Roosevelt

Friends of the Earth
Northern Great Plains Region
P.O. Box 882
Billings, MT 59103
  (406) 252-3988
  Representative: Ed Dobson

Friends of the Earth
Northern Rockies Region
P.O. Box 496
Yellowsprings, OH 45387
  (513) 767-7331
  Representative: Randall Gloege

Friends of the Earth
Northwest Region
4512 University Way, N.E.
Seattle, WA 98105
  (206) 633-1661
  Representative: Dale Jones

Friends of the Earth
Sacramento Region
Room 209
717 K Street
Sacramento, CA 95804
  (916) 466-3109
  Representative: Michael Storper

Friends of the Earth
Southwest Region
P.O. Box 131
Route 3
Santa Fe, NM 87501
  (505) 982-0953
  Representative: Sally Rodgers

Fullerton City
Planning Department
City Hall
Fullerton, CA 92632
  (714) 525-7171
  Director

Fulton County
Planning Department
County Courthouse

Rochester, IN 46975
  (219) 223-2911
  Director

Fulton County
Planning Department
County Courthouse
Johnstown, NY 12095
  (518) 762-4128
  Director: Paul O'Connor

Gallatin County
Planning Department
County Courthouse
Bozeman, MT 59715
  (406) 587-4271
  EIS Contact: Stuart Westlake

Garden Grove City
Planning Department
City Hall
Garden Grove, CA 92640
  (714) 638-6623
  Director

Garfield County
Planning Department
County Courthouse
Jordan, MT 59337
  (406) 557-2760
  Director

Garfield County
Planning Department
County Courthouse
Pomeroy, WA 99347
  (509) 843-1391
  Director

Garrett County
Planning Office
323 East Oak Street
Oakland, MD 21550
  (301) 334-4200
  EIS Contact: Tim Dugan

Gary City
Planning Department
City Hall
Gary, IN 46402
  (219) 944-1500
  Director

Gaston County
Planning Department
P.O. Box 1578
Gastonia, NC 28052
  (704) 865-6411
  Director

Gates County
Planning Department
County Courthouse
Gatesville, NC 27938
  (919) 357-1365
  Director

Genesee County
Planning Department
County Courthouse
Batavia, NY 14020
  (716) 344-2100
  Director

Georgia Association of Conservation
  District Supervisors
Route 1
Blakely, GA 31723
  (912) 723-5074
  President: Ralph Balkcome

Georgia Association of County Commis-
  sioners
Suite 1120
Carnegie Building
Atlanta, GA 30303
  (404) 522-5022
  Executive Director: Hill Healan

Georgia Conservancy
3110 Maple Drive
Suite 407
Atlanta, GA 30305
  (404) 262-1967
  Chairman: Donald Downing

Georgia
Department of Natural Resources
270 Washington Street, S.W.
Atlanta, GA 30334
  (404) 656-3530
  Commissioner: Joe Tanner

Georgia
Department of Natural Resources

Environmental Protection Division
270 Washington Street, S.W.
Atlanta, GA   30334
   (404) 656-3530
   Director: Leonard Ledbetter

Georgia Environmental Council
3110 Maple Drive
Suite 407
Atlanta, GA   30305
   (404) 993-7124
   President: Betsy Loyless

Georgia Institute of Natural Resources
University of Georgia
Room 13
Ecology Building
Athens, GA   30602
   (404) 542-1555
   Director: Ronald North

Georgia Municipal Association
220-10 Pryor Street Building
Atlanta, GA   30303
   (404) 688-0472
   Executive Director: Elmer George

Georgia State Soil and Water Conservation Committee
1867 West Broad Street
Athens, GA   30606
   (404) 542-3065
   Acting Executive Director: Dennis
      Hopper

Gibson County
Planning Department
County Courthouse
Princeton, IN   47570
   (812) 385-5529
   Director

Giles County
Planning Department
507 Wenonah Avenue
Pearisburg, VA   24134
   (703) 921-2525
   Director

Glacier County
Planning Department
County Courthouse

Cut Bank, MT   59427
   (406) 873-2482
   Director

Glendale City
Planning Department
613 East Broadway
Glendale, CA   91205
   (213) 956-4000
   Director

Glenn County
Planning Department
County Courthouse
Willows, CA   95988
   (916) 934-3364
   Director

Gloucester County
Planning Department
P.O. Box 329
Gloucester, VA   23061
   (804) 693-4042
   Director

Golden Valley County
Planning Department
County Courthouse
Ryegate, MT   59074
   (406) 568-2231
   Director

Goochland County
Planning Department
County Courthouse
Goochland, VA   23063
   (804) 556-4701
   Director

Goodhue County
Planning Department
County Courthouse
Red Wing, MN   55066
   (612) 388-8261
   Director

Govenor's Commission on Arizona
   Environment
206 South 17th Avenue
Phoenix, AZ   85007
   (602) 261-7803
   Chairman: Roy Drachman

Graham County
Planning Department
County Courthouse
Robbinsville, NC 28771
  (704) 479-3411
  Director

Granite County
Planning Department
County Courthouse
Philipsburg, MT 59858
  (406) 859-3771
  Director

Grant County
Planning Department
428 South Washington Street
Marion, IN 46952
  (317) 668-8121
  EIS Contact: Betty Pence

Grant County
Planning Department
County Courthouse
Elbow Lake, MN 56531
  (218) 685-4520
  Director

Grant County
Planning Department
County Courthouse
Milbank, SD 57252
  (605) 432-5482
  Director

Grant County
Planning Department
County Courthouse
Ephrata, WA 98823
  (509) 754-2011
  Director: W.H. Henager

Grant County
Planning Department
County Courthouse
Lancaster, WI 53813
  (608) 723-2675
  Director

Grant Lincoln Adams Conference of
  Governments
c/o Grant County Planning Office

Grant County Courthouse
Ephrata, WA 98823
  (509) 754-2011
  Director

Granville County
Planning Department
County Courthouse
Oxford, NC 27565
  (919) 693-2649
  Director

Grays Harbor County
Planning Department
County Courthouse
Montesano, WA 98563
  (206) 249-3731
  Director

Grays Harbor Regional Planning
  Commission
207½ East Market Street
Aberdeen, WA 98520
  (206) 532-8812
  Director

Grayson County
Planning Department
Box 358
Independence, VA 24348
  (703) 773-7302
  Director

Greater Bridgeport Regional Planning
  Agency
Bridgeport Transportation Center
525 Water Street
Bridgeport, CT 06604
  (203) 366-5405
  Director

Great Falls City
Planning Department
City Hall
Great Falls, MT 59401
  (406) 727-5881
  Director

Great Falls City-County Planning
  Board
P.O. Box 1609
Great Falls, MT 59403

(406) 727-5881
Director

Great Lakes Basin Commission
P.O. Box 999
Ann Arbor, MI 48106
  (313) 668-2340
  Program Manager: Charles Job

Great Lakes Tomorrow
53 West Jackson Boulevard
Chicago, IL 60604
  (312) 427-5123
  President: Mimi Becker

Green Bay City
Planning Department
City Hall
Green Bay, WI 54301
  (414) 497-3622
  Director

Green County
Planning Department
County Courthouse
Monroe, WI 53566
  (608) 328-8288
  Director

Greene County
Planning Department
County Courthouse
Bloomfield, IN 47424
  (812) 384-8532
  Director

Greene County
Planning Department
County Courthouse
Catskill, NY 12414
  (518) 943-2050
  Director

Greene County
Finance Office
P.O. Box 5
Snow Hill, NC 28580
  (919) 747-3505
  Finance Officer: G.L. Mewborn, Jr.

Greene County
Planning Department

P.O. Box 358
Standardsville, VA 22973
  (804) 985-7803
  Director

Green Lake County
Planning Department
County Courthouse
Green Lake, WI 54941
  (414) 294-6581
  Director

Greensboro City
Planning Department
City Hall
Greensboro, NC 27402
  (919) 373-2397
  EIS Contact: Dan Curry

Greensville County
Planning Department
301 South Main Street
P.O. Box 109
Emporia, VA 23847
  (804) 634-2038
  Director

Greenwich Town
Planning Department
Town Hall
Greenwich, CT 06830
  (203) 622-7700
  Director

Gregory County
Planning Department
County Courthouse
Burke, SD 57523
  (605) 775-2665
  Director

Guam
Environmental Protection Agency
Government of Guam
P.O. Box 2999
Agana, GU 96910
  Overseas Operator 646-7916
  Administrator: O.V. Natarajan

Guilford County
Planning Department
P.O. Box 3427

Greensboro, NC 27402
  (919) 373-2000
  Director

Haakon County
Planning Department
County Courthouse
Philip, SD 57567
  (605) 859-2800
  Director

Halifax County
Planning Department
County Courthouse
Halifax, NC 27839
  (919) 583-5061
  Director

Halifax County
Planning Department
P.O. Box 786
Halifax, VA 24558
  (804) 476-2141
  Director

Hamilton County
Planning Department
County Courthouse
Noblesville, IN 46060
  (317) 773-6110
  Director

Hamilton County
Planning Department
County Courthouse
Lake Pleasant, NY 12108
  (518) 548-7111
  Director

Hamlin County
Planning Department
County Courthouse
Hayti, SD 57241
  (605) 783-3201
  Director

Hammond City
Planning Department
7324 Indianapolis Boulevard
Hammond, IN 46324
  (219) 853-6300
  EIS Contact: David Thomas

Hampden County
Planning Department
County Courthouse
Springfield, MA 01101
  (413) 781-8100
  Director

Hampshire County
Planning Department
County Courthouse
Northampton, MA 01060
  (413) 584-0557
  Director

Hampton City
Planning Department
City Hall
Hampton, VA 23369
  (804) 727-6000
  Director

Hancock County
Planning Commission
County Courthouse
First Floor
Greenfield, IN 46140
  (317) 462-7569
  EIS Contact: Noble Snodgrass

Hand County
Planning Department
County Courthouse
Miller, SD 57362
  (605) 853-3337
  Director

Hanover County
Planning Department
County Courthouse
Hanover, VA 23069
  (804) 798-6081
  Director: York Phillips

Hanson County
Planning Department
County Courthouse
Alexandria, SD 57311
  (605) 239-4446
  Director

Harding County
Planning Department

County Courthouse
Buffalo, SD 57720
   (605) 375-3351
   Director

Harford County
Planning Department
45 South Main Street
Bel Air, MD 21014
   (301) 838-6000
   Director

Harnett County
Planning Department
County Courthouse
Lillington, NC 27546
   (919) 893-2654
   Director

Harrison County
Planning Department
County Courthouse
Corydon, IN 47112
   (812) 738-8149
   Director

Hartford City
Planning Department
City Hall
Hartford, CT 06103
   (203) 566-6400
   Director

Hawaii Association of Conservation
   Districts
P.O. Box 537
Pahala, HI 96777
   (808) 928-8311
   President: Jay Sasan

Hawaii County
Planning Department
25 Aupuni Street
Hilo, HI 96720
   (808) 961-8222
   Director: Sidney Fuke

Hawaii
Department of Land and Natural
   Resources
Box 621
Honolulu, HI 96809

   (808) 548-6550
   Chairman: William Thompson

Hawaii
Department of Land and Natural
   Resources
Division of Land Management
P.O. Box 621
Honolulu, HI 96809
   (808) 548-7515
   Administrator: James Detor

Hawaii Environmental Center
University of Hawaii
2550 Campus Road
Honolulu, HI 96822
   (808) 948-7361
   Director: Dr. Doak Cox

Hawaii
Environmental Quality Commission
550 Halekauwila Street
Honolulu, HI 96813
   (808) 548-6915
   Executive Secretary

Hawaii
Office of Environmental Quality
   Control
550 Halekauwila Street
Honolulu, HI 96813
   (808) 548-6915
   Director: Richard O'Connell

Hawaii
State Association of Counties
25 Aupuni Street
Hilo, HI 96720
   (808) 961-8267
   President: Tomio Fujii

Hawthorne City
Planning Department
4455 West 126th Street
Hawthorne, CA 90250
   (213) 676-1181
   Director: Bradley Stevens

Hayward City
Planning Department
22300 Foothill Boulevard
Hayward, CA 94541

(415) 581-2345
Senior Planner: Marvin Carash

Haywood County
Planning Department
County Courthouse
Waynesville, NC 28786
   (704) 456-9812
   Director

Headwaters Regional Development
   Commission
15th and Delton
Box 584
Bemidji, MN 56601
   (218) 751-3108
   Director

Helena Lewis and Clark City-County
   Planning Department
38 South Last Chance Gulch
Helena, MT 59601
   (406) 442-5000
   Director

Henderson County
Planning Department
244 Second Avenue, East
Hendersonville, NC 28739
   (704) 692-4213
   EIS Contact: David Fowler

Hendricks County
Planning Department
County Courthouse
Danville, IN 46122
   (317) 745-2794
   Director

Hennepin County
Planning Department
2300 Government Center
Minneapolis, MN 55487
   (612) 348-3000
   Director

Henrico County
Planning Department
P.O. Box 27032
Richmond, VA 23273
   (804) 747-4602
   Principal Planner: Angela Moore

Henry County
Planning Department
County Courthouse
New Castle, IN 47362
   (317) 529-2800
   Director

Henry County
Planning Department
County Courthouse
Martinsville, VA 24112
   (703) 638-5311
   Director

Herkimer County
Planning Department
County Office Building
Herkimer, NY 13350
   (315) 866-4010
   Director

Herkimer-Oneida Counties Comprehensive Planning Program
Oneida County Office Building
800 Park Avenue
Utica, NY 13501
   (315) 798-5718
   Principal Planner: John Kent

Hertford County
Planning Department
County Courthouse
Winton, NC 27986
   (919) 358-4271
   Director

Highland County
Planning Department
County Courthouse
Monterey, VA 24465
   (703) 468-2447
   Director

High Plains Council for District One
P.O. Box 749
Scobey, MT 59263
   (406) 487-2241
   Director

High Point City
Planning Department
P.O. Box 230

High Point, NC 27261
  (919) 887-2511
  EIS Contact: Reggie Greenwood

Hill County
Planning Department
County Courthouse
Havre, MT 59501
  (406) 265-5481
  Director

Hoke County
Planning Department
County Courthouse
Raeford, NC 28376
  (919) 875-2034
  Director

Holyoke City
Planning Department
City Hall
Holyoke, MA 01040
  (413) 532-2602
  Director

Honolulu City
Planning Department
650 South King Street
Honolulu, HI 96813
  (808) 523-4111
  EIS Contact: Lon Polk

Honolulu County
Planning Department
County Courthouse
Honolulu, HI 96813
  (808) 523-4111
  Director

Housatonic Valley Council of Elected
  Officials
256 Main Street
Danbury, CT 06810
  (203) 743-2769
  Director

Houston County
Planning Department
County Courthouse
Caledonia, MN 55921
  (507) 724-3930
  EIS Contact: Charles Sheehan

Howard County
Planning Commission
County Courthouse
Kokomo, IN 46901
  (317) 459-3131
  Director: James Daily

Howard County
Department of Public Works
3430 Courthouse Drive
Ellicott City, MD 21043
  (301) 992-2400
  EIS Contact: John O'Hara

Hubbard County
Planning Department
County Courthouse
Park Rapids, MN 56470
  (218) 732-5286
  Director

Hudson Valley Regional Council
c/o County Executive Office
Orange County Government Center
Goshen, NY 10924
  (914) 294-8244
  Director

Hughes County
Planning Department
County Courthouse
Pierre, SD 57501
  (605) 773-3713
  Director

Humboldt County
Planning Department
825 Fifth Street
Eureka, CA 95501
  (707) 445-7266
  Director

Humboldt County Association of
  Governments
P.O. Box 1018
Eureka, CA 95501
  (707) 443-7331
  Director

Huntington Beach City
Planning Department
P.O. Box 190

Huntington Beach, CA   92648
    (714) 536-5511
    Director

Huntington County
Planning Department
County Courthouse
Huntington, IN   46750
    (219) 356-7618
    Director

Hutchinson County
Planning Department
County Courthouse
Olivet, SD   57052
    (605) 387-5335
    Director

Hyde County
Planning Department
County Courthouse
Swanquarter, NC   27885
    (919) 926-4101
    Director

Hyde County
Planning Department
County Courthouse
Highmore, SD   57345
    (605) 852-2512
    Director

Idaho Association of Cities
3314 Grace Street
Boise, ID   83703
    (208) 344-8594
    Martin Peterson

Idaho Association of Counties
P.O. Box 1623
Boise, ID   83701
    (208) 345-9126
    Executive Director: Colen Sweeten

Idaho
Department of Health and Welfare
Division of the Environment
Statehouse
Boise, ID   83720
    (208) 964-3109
    Director: Dr. Lee Stokes

Idaho Environmental Council
P.O. Box 1708
Idaho Falls, ID   83401
    (208) 523-6692
    President: Gerald Jayne

Illinois Association of County Board
    Members
403 West Edwards Street
Springfield, IL   62704
    (217) 528-5331
    Executive Secretary: Paul
        Bitschenauer

Illinois Association of Soil and Water
    Conservation Districts
RR 2
Polo, IL   61604
    (815) 946-2301
    President: Wilbur Bowman

Illinois Counties Council
105 West Adams Street
30th Floor
Chicago, IL   60603
    (312) 346-7500
    Executive Director: Arthur
        Gottschalk

Illinois Environmental Council
407½ East Adams Street
Springfield, IL   62701
    (217) 544-5954
    Director: Sandra McAvoy

Illinois
Environmental Protection Agency
2200 Churchill Road
Springfield, IL   62706
    (217) 782-3397
    Director: Michael Mauzy

Illinois Municipal League
1220 South 7th Street
Springfield, IL   62703
    (217) 525-1220
    Executive Director: Steven Sargent

Imperial County
Planning Department
County Courthouse
El Centro, CA   92243

(714) 352-3610
EIS Contact: Sari McClure

Indiana Association of Cities and
   Towns
408 ISTA Center
150 West Market Street
Indianapolis, IN 46204
   (317) 635-8616
   Executive Director: Michael Quinn

Indiana Association of Counties
317 Illinois Building
17 West Market Street
Indianapolis, IN 46204
   (317) 632-7453
   Executive Director: Shirl Evans, Jr.

Indiana Association of Soil and Water
   Conservation Districts
17414 Comer Road
Fort Wayne, IN 46819
   (219) 639-6147
   President: Ellis McFadden

Indiana Conservation Council
P.O. Box 672
Muncie, IN 47305
   (317) 288-0368
   President: Claude Ferguson

Indiana
Department of Natural Resources
608 State Office Building
Indianapolis, IN 46204
   (317) 633-6344
   Director: Joseph Cloud

Indiana
Department of Natural Resources
State Soil and Water Conservation
   Committee
Room 7
AGAD Building
Purdue University
West Lafayette, IN 47907
   (317) 749-2364
   Executive Secretary: Charles McKee

Indiana Heartland Coordinating
   Commission
7212 North Shadeland Avenue

Indianapolis, IN 46250
   (317) 849-4629
   Director

Indianapolis City
Planning Department
City Hall
Indianapolis, IN 46204
   (317) 633-3200
   Director

Inglewood City
Planning Department
One Manchester Boulevard
P.O. Box 6500
Inglewood, CA 90301
   (213) 649-7280
   EIS Contact: Karen Heit

Institute of Environmental Sciences
940 East Northwest Highway
Mt. Prospect, IL 60056
   (312) 255-1561
   Director

Institute on Man and Science
Economic and Environmental Studies
Pond Hill Road
Rensselaerville, NY 12147
   (518) 797-3783
   Dr. Gordon Enk

International Environmental Resources
   Network
P.O. Box 417
Concord, MA 01742
   (617) 969-7100
   Director: John Whitman

Interstate Commission on the Potomac
   River Basin
4350 East-West Highway
Suite 814
Bethesda, MD 20014
   (301) 652-5758
   Executive Director: Paul Eastman

Inyo County
Planning Department
P.O. Drawer N
Independence, CA 93526
   (714) 878-2411

Director

**Inyo Mono Association of Government
   Entities**
301 West Line Street
Suite G
Bishop, CA 93514
   (714) 878-2411
   EIS Contact: Randy Pestor

**Iowa Citizens for Environmental
   Quality**
P.O. Box 1147
Ames, IA 50010
   (515) 432-5943
   Chairman: James O'Toole

Iowa County
Planning Department
County Courthouse
Dodgeville, WI 53533
   (608) 935-5445
   Director

Iowa
Department of Environmental Quality
Wallace Building
900 East Grand
Des Moines, IA 50319
   (515) 265-8854
   Executive Director: Larry Crane

Iowa
Department of Environmental Quality
Wallace Building
900 East Grand
Des Moines, IA 50319
   (515) 281-8884
   Environmental Specialist: John Seyb

Iowa League of Municipalities
Suite 100
900 Des Moines Street
Des Moines, IA 50316
   (515) 265-9961
   Executive Director: Robert Harpster

Iowa State Association of Counties
730 East Fourth Street
Des Moines, IA 50316
   (515) 244-7181

Executive Director: Donald
   Cleveland

Iredell County
Planning Department
P.O. Box 803
Statesville, NC 28677
   (704) 872-9821
   EIS Contact: Lisa Beckham

Iron County
Planning Department
County Courthouse
Hurley, WI 54534
   (715) 561-3375
   Director

Irondequoit City
Planning Department
City Hall
Irondequoit, NY 14617
   (716) 467-8840
   Director

Isanti County
Planning Department
County Courthouse
Cambridge, MN 55008
   (612) 689-2292
   Director

Island County
Planning Department
County Courthouse
Coupeville, WA 98239
   (206) 678-5111
   Director

Isle of Wight County
Planning Department
P.O. Box 268
Isle of Wight, VA 23487
   (804) 357-3191
   Director: Terry Lewis

**Isothermal Planning and Development
   Commission**
Box 841
Rutherfordton, NC 28139
   (704) 652-8098
   Director

Itasca County
Planning Department
County Courthouse
Grand Rapids, MN 55744
  (218) 326-9777
  Director

Izak Walton League of America
1800 North Kent Street
Suite 806
Arlington, VA 22209
  (703) 528-1818
  President: Howard White

Jackson County
Planning Department
County Courthouse
Brownstown, IN 47220
  (812) 358-4242
  Director

Jackson County
Planning Department
County Courthouse
Jackson, MN 56143
  (507) 847-2763
  Director

Jackson County
Planning Department
County Courthouse
Sylva, NC 28779
  (704) 586-4312
  Director

Jackson County
County Auditor
Kadoka, SD 57543
  (605) 837-2422
  Vicki Wilson

Jackson County
Planning Department
County Courthouse
Black River Falls, WI 54615
  (715) 284-2221
  Zoning Administrator: Steve Raith

James City County
Planning Department
P.O. Box JC
Williamsburg, VA 23185

  (804) 220-1122
  Director

Jasper County
Planning Department
County Courthouse
Rensselaer, IN 47978
  (219) 866-7421
  Director

Jay County
Planning Department
County Courthouse
Portland, IN 47371
  (219) 726-7595
  Director

Jefferson County
Planning Department
County Courthouse
Madison, IN 47250

Jefferson County
Planning Department
County Courthouse
Watertown, NY 13601
  (315) 785-3081
  Director

Jefferson County
Planning Department
County Courthouse
Boulder, MT 59632
  (406) 225-3332
  Director

Jefferson County
Planning Department
County Courthouse
Port Townsend, WA 98368
  (206) 385-1427
  Director

Jefferson County
Planning Department
County Courthouse
Jefferson, WI 53549
  (414) 674-2500
  Director

Jefferson-Port Townsend Regional
  Council Planning Department

Courthouse
Port Townsend, WA 98368
  (206) 385-1427
  Director

Jennings County
Planning Department
County Courthouse
Vernon, IN 47282
  (812) 346-5977
  Director

Jerauld County
Planning Department
County Courthouse
Wessington Springs, SD 57382
  (605) 539-1202
  Director

John Muir Institute for Environmental
  Studies
743 Wilson Street
Napa, CA 94558
  (707) 252-8333
  President: Max Linn

Johnson County
Planning Department
County Courthouse
Franklin, IN 46131
  (317) 736-9090
  Director

Johnston County
Planning Department
County Courthouse
Smithfield, NC 27577
  (919) 934-3191
  Director

Jones County
Planning Department
P.O. Box 266
Trenton, NC 28585
  (919) 448-2571
  Director

Jones County
Planning Department
County Courthouse
Murdo, SD 57559
  (605) 669-2242

Director

Judith Basin County
Planning Department
County Courthouse
Stanford, MT 59479
  (406) 566-2301
  Director

Juneau County
Planning Department
Courthouse Annex
Room 26
Mauston, WI 53948
  (608) 847-4690
  EIS Contact: Dennis Emery

Kalispell City-County Planning Board
723 Fifth Avenue East
Kalispell, MT 59901
  (406) 755-5300
  Director

Kanabec County
Planning Department
County Courthouse
Mora, MN 55051
  (612) 679-1030
  Director

Kandiyohi County
Planning Department
County Courthouse
Willmar, MN 56201
  (612) 235-2727
  Director

Kankakee-Iroquois Regional Planning
  Commission
P.O. Box 708
Francesville, IN 47946
  (219) 567-9432
  EIS Contact: Chris Larson

Kansas Association of Conservation
  Districts
Munjor Tr.
Hays, KS 67601
  (913) 625-5430
  President: Robert Binder

Kansas Association of Counties

112 West Seventh Street
Topeka, KS 66603
  (913) 233-2271
  Executive Secretary: Fred Allen

Kansas League of Municipalities
112 West 7th Street
Topeka, KS 66603
  (913) 345-9565
  Executive Director: E.A. Mosher

Kansas
State Department of Health and
  Environment
Division of Environment
Forbes Field Building 740
Topeka, KS 66620
  (913) 862-9360
  Director: Melville Gray

Kauai County
Planning Department
4396 Rice Street
Lihue, HI 96766
  (808) 245-4785
  Director

Kenosha City
Planning Department
City Hall
Kenosha, WI 53140
  (414) 656-6000
  Director

Kenosha County
Planning Department
County Courthouse
Kenosha, WI 53140
  (414) 656-6400
  Director

Kent County
Planning Department
County Courthouse
Chestertown, MD 21620
  (301) 778-4600
  Director

**Kentucky Association of Conservation
  Districts**
Rt. 1
Robards, KY 42452

  (502) 521-7532
  President: George Crafton

Kentucky Association of Counties
P.O. Box 345
Frankfort, KY 40601
  (502) 223-7668
  Executive Director: Fred Creasey

Kentucky
Department for Natural Resources and
  Environmental Protection
5th Floor
Capital Plaza Tower
Frankfort, KY 40601
  (502) 564-3350
  Commissioner: A.L. Roark

Kentucky Municipal League
202 Bradley Hall
University of Kentucky
Lexington, KY 40506
  (606) 257-2785
  Executive Director: Michael Amyx

Kern County
Planning Department
1103 Golden State Avenue
Bakersfield, CA 93301
  (805) 861-2618
  Principal Planner: Frederick Simon

Kern County Council of Governments
1106-26th Street
Bakersfield, CA 93301
  (805) 861-2191
  Assistant Director: Bradley Williams

**Kerr-Tar Regional Council of Govern-
  ments**
P.O. Box 709
238 Orange Street
Henderson, NC 27536
  (919) 492-8561
  Director

Kewaunee County
Planning Department
County Courthouse
Kewaunee, WI 54216
  (414) 338-3580
  Director

Kickitat County
Planning Department
County Courthouse
Goldendale, WA 98620
   (509) 773-4612
   Director

King and Queen County
Planning Department
County Courthouse
King and Queen Court House, VA   23085
   (804) 785-7955
   Director

King County
Planning Department
450 Administration Building
Seattle, WA 98104
   (206) 344-7900
   EIS Contact: David Feltman

Kingsbury County
Planning Department
County Courthouse
De Smet, SD 57231
   (605) 854-3811
   EIS Contact: Alfred Schoenfelder

Kings County
Planning Agency
County Courthouse
Hanford, CA 93230
   (209) 582-3211
   Deputy Director: William Zumwalt

Kings County Regional Planning Agency
Courthouse
Box C
Hanford, CA 93230
   (209) 582-3211
   Director

King George County
Planning Department
P.O. Box 198
King George, VA 22485
   (703) 775-9181
   County Administrator: Ken Scruggs

King William County
Planning Department
County Courthouse

King William, VA   23066
   (804) 769-2671
   Director

Kitsap County
Planning Department
County Courthouse
614 Division Street
Port Orchard, WA 98366
   (206) 876-7181
   EIS Contact: Rick Kimball

Kittitas County
Planning Department
Room 216
Courthouse
Ellensburg, WA 98926
   (509) 925-4631
   Director: Tom Pickerel

Kittitas County Conference of Govern-
   ments
Kittitas County Courthouse
Ellensburg, WA 98926
   (509) 925-4631
   Director

Kittson County
Planning Department
County Courthouse
Hallock, MN 56728
   (218) 843-3801
   Director

Klickitat Regional Council
P.O. Box 268
Goldendale, WA 98620
   (509) 773-5703
   Director

Knox County
Area Plan Commission
County Courthouse
Vincennes, IN 47591
   (812) 882-6384
   Executive Director: Glenn Koby

Koochiching County
Planning Department
County Courthouse
International Falls, MN 56649
   (218) 283-2583

EIS Contact: Thomas Barthell

Kosciusko County
Planning Department
County Courthouse
Warsaw, IN 46580
 (219) 267-4444
 Director

Lac Qui Parle County
Planning Department
County Courthouse
Madison, MN 56256
 (612) 598-7444
 Director

La Crosse City
Planning Department
P.O. Box 945
La Crosse, WI 54601
 (608) 782-5655
 Director

La Crosse County
Planning Department
County Courthouse
La Crosse, WI 54601
 (608) 785-9581
 Director

Lafayette County
Planning Department
County Courthouse
Darlington, WI 53530
 (608) 776-4003
 Director

Lagrange County
Planning Department
County Courthouse
Lagrange, IN 46761
 (219) 463-3442
 Director

Lake Champlain-Lake George Regional
 Planning Board
Lake George Institute
Lake George, NY 12845
 (518) 668-5773
 Director

Lake County-City Areawide Planning

255 North Forbes Street
Lakeport, CA 95453
 (707) 263-2211
 Director

Lake County
Planning Department
255 North Forbes Street
Lakeport, CA 95453
 (707) 263-2371
 Environmental Administrator

Lake County
Planning Department
County Courthouse
Crown Point, IN 46307
 (219) 663-0760
 Director

Lake County
Planning Department
County Courthouse
Two Harbors, MN 55616
 (218) 834-4393
 Director

Lake County
Planning Department
County Courthouse
Poison, MT 59860
 (406) 883-4361
 Director

Lake County
Planning Department
County Courthouse
Madison, SD 57042
 (605) 256-2068
 Director

Lake of the Woods County
Planning Department
County Courthouse
Baudette, MN 56623
 (218) 634-1451
 Director

Lakewood City
Planning Department
P.O. Box 158
5050 North Clark Avenue
Lakewood, CA 90714

(213) 866-9771
Assistant Director

Lancaster County
Planning Department
County Courthouse
Lancaster, VA 22503
  (804) 462-2481
  Director

Land-of-Sky Regional Council
P.O. Box 2175
Asheville, NC 28802
  (704) 254-8131
  Director

Langlade County
Planning Department
County Courthouse
Antigo, WI 54409
  (715) 623-3305
  Director

La Porte County
Planning Department
Courthouse Square
La Porte, IN 46350
  (219) 362-7061
  County Planner: Carl Baxmeyer

Lassen County
Planning Department
County Courthouse
Susanville, CA 96130
  (916) 257-5126
  Director

Lawrence City
Planning Department
City Hall
Lawrence, MA 01840
  (617) 686-6177
  Director

Lawrence County
Planning Department
County Courthouse
Bedford, IN 47421
  (812) 275-7543
  Director

Lawrence County

Planning Department
County Courthouse
Deadwood, SD 57738
  (605) 578-1941
  Director

League of Women Voters
Environmental Quality Department
1730 M Street, N.W.
Washington, DC 20036
  (202) 296-1770
  Coordinator: Carol Jolly

Lee County
Planning Department
P.O. Box 987
Sanford, NC 27330
  (919) 775-2103
  Director

Lee County
Planning Department
County Courthouse
Jonesville, VA 24263
  (703) 346-1741
  Director

Lenoir County
Planning Department
County Courthouse
Kinston, NC 28501
  (919) 527-6231
  Director

Lenowisco Planning District Commission
US 58-421 West
Duffield, VA 24244
  (703) 431-2206
  Director: Ronald Flanary

Le Sueur County
Planning Commission
County Courthouse
Le Center, MN 56057
  (612) 357-2251
  EIS Contact: Terry Overn

Lewis and Clark County
Planning Department
County Courthouse
Helena, MT 59601
  (406) 442-6330

Director

Lewis County
Planning Department
County Courthouse
Lowville, NY 13367
  (315) 376-2414
  Director

Lewis County
Planning Department
P.O. Box 418
Chehalis, WA 98532
  (206) 748-9121
  Director: Michael Zengel

Lewis Regional Planning Commission
P.O. Box 418
Chehalis, WA 98532
  (206) 748-6638
  Director

Lewiston City-County Planning Board
P.O. Box 739
Lewiston, MT 59457
  (406) 538-9046
  Director: John Hughes

Liberty County
Planning Department
County Courthouse
Chester, MT 59522
  (406) 759-5365
  Director

Life of the Land
404 Piikoi Street
Room 209
Honolulu, HI 96814
  (808) 521-1300
  President: E. Cooper Brown

Lincoln County
Planning Department
County Courthouse
Ivanhoe, MN 56142
  (507) 694-1529
  Director

Lincoln County
Planning Department
County Courthouse

Libby, MT 59923
  (406) 293-7781
  Director

Lincoln County
Planning Department
County Courthouse
Lincolnton, NC 28092
  (704) 735-6510
  Director

Lincoln County
Planning Department
County Courthouse
Canton, SD 57013
  (605) 987-5891
  Director

Lincoln County
Planning Department
County Courthouse
Davenport, WA 99122
  (509) 725-3031
  Director

Lincoln County
Planning Department
County Courthouse
Merrill, WI 54452
  (715) 536-7444
  Director

Litchfield Hills Regional Planning
  Agency
40 Main Street
Torrington, CT 06790
  (203) 482-5575
  Director

Livingston County
Planning Board
Building Number 2
Murray Hill
Mt. Morris, NY 14510
  (716) 243-2500
  Director

Long Beach City
Planning Department
205 West Broadway
Long Beach, CA 90802
  (213) 590-6075

Director

Lord Fairfax Planning District
  Commission
103 East Sixth Street
Fort Royal, VA 22630
  (703) 635-4146
  Director

Los Angeles City
Planning Department
City Hall
Los Angeles, CA 90012
  (213) 485-2121
  Director

Los Angeles County
Planning Department
320 West Temple Street
Los Angeles, CA 90012
  (213) 974-6469
  EIS Contact: Raymond Ristie

Loudoun County
Planning Department
18 North King Street
Leesburg, VA 22075
  (703) 777-0200
  Environmental Planner: Richard Rein

Louisa County
Planning Department
County Courthouse
Louisa, VA 23093
  (703) 967-0401
  Director

Louisiana Association of Conservation
  Districts
P.O. Box 95
Opelousas, LA 70570
  (318) 942-5639
  President: Richard Hollier

Louisiana
Department of Health and Human
  Resources
Office of Health Services and En-
  vironmental Quality
P.O. Box 60630
New Orleans, LA 70160
  (504) 568-5050

Acting Assistant Secretary: Dr.
  Hamrick

Louisiana
Department of Natural Resources
Environmental Relations
P.O. Box 1628
Baton Rouge, LA 70821
  (502) 925-4501
  Chief: Vernon Robinson

Louisiana Municipal Association
Suite 3-B
5615 Corporate Boulevard
Baton Rouge, LA 70808
  (504) 923-3950
  Executive Director: Charles Pasqua

Louisiana Policy Jury Association
1401 Foss Drive
Baton Rouge, LA 70802
  (504) 343-2835
  Executive Secretary: James Hays

Lowell City
Planning Department
City Hall
Lowell, MA 01852
  (617) 454-8821
  Director

Lower Pioneer Valley Regional Planning Commission
26 Central Street
West Springfield, MA 01089
  (413) 781-6045
  Senior Planner: Elizabeth Kidder

Lumber River Council of Governments
P.O. Drawer 1528
Lumberton, NC 28358
  (919) 738-8104
  EIS Contact: Victor Josephs

Luneberg County
Planning Department
County Courthouse
Luneberg, VA 23952
  (804) 696-2230
  Director

Lyman County

Planning Department
County Courthouse
Kennebec, SD 57544
  (605) 869-2277
  Director

Lynchburg City
Planning Department
P.O. Box 60
Lynchburg, VA 24505
  (804) 847-1400
  Director

Lynn City
Planning Department
City Hall
Lynn, MA 01901
  (617) 598-4000
  Director

Lyon County
Planning Department
County Courthouse
Marshall, MN 56258
  (507) 532-2631
  Director

Macon County
Planning Department
County Courthouse
Franklin, NC 28734
  (704) 524-6421
  Director

Madera County
Planning Department
135 West Yosemite Avenue
Madera, CA 93637
  (209) 674-4641
  Planner: Peggy Palms

Madison City
Planning Department
Room 414
City-County Building
Madison, WI 53709
  (608) 266-4635
  EIS Contact: David Larson

Madison County
Planning Department
County Courthouse

Anderson, IN 46016
  (317) 646-9212
  Director

Madison County
Planning Department
County Courthouse
Virginia City, MT 59755
  (406) 843-5311
  Director

Madison County
Planning Board
Box 606
Wampsville, NY 13163
  (315) 366-2376
  Director

Madison County
Planning Department
County Courthouse
Marshall, NC 28753
  (704) 649-2521
  Director

Madison County
Planning Department
P.O. Box 220
Madison, VA 22727
  (703) 948-4561
  Director

Madison County Council of Governments
Government Center
Anderson, IN 46016
  (317) 646-9338
  Director

Mahnomen County
Planning Department
County Courthouse
Mahnomen, MN 56557
  (218) 935-5742
  Director

Maine Association of Conservation
  Commissions
Box 548
Kennebunkport, ME 04046
  (207) 967-3705
  President: Carl Laws

Maine Association of Conservation
  Districts
R.D.
Fort Fairfield, ME   04742
  (207) 476-5032
  President: Louis McLaughlin

Maine Coast Heritage Trust
60 Main Street
Bar Harbor, ME   04609
  (207) 288-5019
  President: Harold Woodsum, Jr.

Maine County Commissioners Association
2 Turner Street
Auburn, ME   04210
  (207) 782-6131
  Executive Secretary: Roland Landry

Maine
Department of Conservation
Administrative Services
State Office Building
Augusta, ME   04333
  (207) 289-2212
  Director: A. Temple Bowen, Jr.

Maine Municipal Association
Local Government Center
Community Drive
Augusta, ME   04330
  (207) 623-8429
  Executive Director: John Salisbury

Malden City
Planning Department
City Hall
Malden, MA   02148
  (617) 324-6600
  Director

Manitowoc County
Planning Department
County Courthouse
Manitowoc, WI   54220
  (414) 682-8811
  Director

Marathon County
Planning Department
County Courthouse
Wausau, WI   54401

  (715) 842-2141
  Director

Marin County
Planning Department
County Courthouse
San Rafael, CA   94903
  (415) 479-1100
  Director

Marine Environmental Council of Long
  Island
P.O. Box 55
Seaford, NY   11783
  (516) 221-1434
  President: Vincent Franco

Marinette County
County Clerk
Marinette, WI   54143
  (715) 735-3371
  Donald John

Marion County
Planning Department
County Courthouse
Indianapolis, IN   46204
  (317) 633-3200
  Director

Mariposa County
Planning Department
County Courthouse
Mariposa, CA   95338
  (209) 966-2005
  Director

Marquette County
Planning Department
P.O. Box 21
Montello, WI   53949
  (414) 297-2136
  Zoning Administrator: Michael
    Stapleton

Marshall County
Planning Department
County Courthouse
Plymouth, IN   46563
  (219) 936-3359
  Director

Marshall County
County Auditor
County Courthouse
Warren, MN 56762
  (218) 745-4921
  EIS Contact: Charles Cheney

Marshall County
Planning Department
County Courthouse
Britton, SD 57430
  (605) 448-5171
  Director

Martha's Vineyard Commission
Box 1447
Oak Bluffs, MA 02557
  (617) 693-3453
  Director

Martin County
Planning Department
County Courthouse
Shoals, IN 47581
  (812) 247-3651
  Director

Martin County
Zoning and Solid Waste Office
Room 124
Security Building
Fairmont, MN 56031
  (507) 238-4757
  Zoning Administrator: Wayne Flohrs

Martin County
Planning Department
County Courthouse
Williamston, NC 27892
  (919) 792-2515
  Director

Maryland Association of Counties
169 Conduit Street
Annapolis, MD 21401
  (301) 268-5884
  Executive Director: Wallace Hutton

Maryland
Department of Natural Resources
Tawes State Office Building
Annapolis, MD 21401

  (301) 269-3041
  Secretary: James Coulter

Maryland
Department of Natural Resources
Environmental Service
Tawes State Office Building
Annapolis, MD 21401
  (301) 269-3351
  Director: Thomas McKewen

Maryland Environmental Trust
Suite 1401
501 St. Paul Place
Baltimore, MD 21202
  (301) 383-4264
  Chairman: King Burnett

Maryland Municipal League
76 Maryland Avenue
Annapolis, MD 21401
  (301) 268-5514
  Executive Director: John Robinson

Maryland Regional Planning Council
701 St. Paul Street
Baltimore, MD 21202
  (301) 777-2158
  Director

Maryland
State Department of Health and Mental
  Hygiene
Environmental Health Administration
P.O. Box 13387
201 West Preston Street
Baltimore, MD 21203
  (301) 383-2740
  Director: Donald Noren

Maryland University Center for Environmental and Estuarine Study
Cambridge, MD 21613
  (301) 228-9250
  Director: Dr. Peter Wagner

Mason County
Planning Department
P.O. Box 400
Shelton, WA 98584
  (206) 426-3222
  Director

Mason Regional Planning Council
P.O. Box 186
Shelton, WA 98584
  (206) 426-1351
  Director

Massachusetts Association of Conser-
  vation Commissions
Lincoln Filene Center
Tuffs University
Medford, MA 02155
  (617) 628-5000
  President: George Wislocki

Massachusetts Association of Conservation Districts
36 Prospect Street
Sherborn, MA 01770
  (617) 653-7491
  President: Richard Darby

Massachusetts County Commissioners'
  and Sheriffs' Association
Main Street
Barnstable, MA 02630
  (617) 362-3552
  President: John Bowes

Massachusetts
Department of Public Works
100 Nashua Street
Boston, MA 02114
  (617) 727-4800
  Environmental Engineer: John Hurley

Massachusetts
Executive Office of Environmental
  Affairs
Department of Environmental Manage-
  ment
100 Cambridge Street
Boston, MA 02202
  (617) 727-3163
  Commissioner: Richard Kendall

Massachusetts
Executive Office of Environmental
  Affairs
Department of Environmental Quality
  Engineering
100 Cambridge Street
Boston, MA 02202

  (617) 727-2690
  Commissioner: David Standley

Massachusetts
Executive Office of Environmental
  Affairs
Department of Metropolitan District
  Commission
20 Sommerset Street
Boston, MA 02108
  (617) 727-5114
  Commissioner: John Snedeker

Massachusetts
Executive Office of Environmental
  Affairs
Environmental Impact Review
100 Cambridge Street
Boston, MA 02202
  (617) 727-9800
  Director: William Hicks

Massachusetts League of Cities and
  Towns
131 Tremont Street
Boston, MA 02108
  (617) 426-7272
  Executive Director: Kennedy Shaw

Mathews County
Planning Department
County Courthouse
Mathews, VA 23109
  (804) 725-7171
  Director

Maui County
Planning Department
County Courthouse
Wailuku, HI 96793
  (808) 244-7825
  Director

McCone County
Planning Department
County Courthouse
Circle, MT 59215
  (406) 485-3505
  Director

McCook County
Planning Department

County Courthouse
Salem, SD   57058
   (605) 425-2791
   Director

McDowell County
Planning Department
County Courthouse
Marion, NC   28752
   (704) 652-7121
   Director

McLeod County
Zoning Administration
County Courthouse
Glencoe, MN   55336
   (612) 864-5551
   EIS Contact: Edwin Homan

McPherson County
Planning Department
County Courthouse
Leola, SD   57456
   (605) 439-3314
   Director

Meade County
Planning Department
County Courthouse
Sturgis, SD   57785
   (605) 347-2360
   Director

Meagher County
Planning Board
County Courthouse
White Sulphur Springs, MT   59645
   (406) 547-3738
   Director: Kathleen Fuller

Mecklenburg County
Planning Department
720 East Fourth Street
Charlotte, NC   28202
   (704) 374-2472
   Director

Mecklenburg County
Planning Department
P.O. Box 307
Boydton, VA   23917
   (804) 738-6488

Director

Medford City
Planning Department
City Hall
Medford, MA   02155
   (617) 396-5500
   Director

Meeker County
Planning Department
County Courthouse
Litchfield, MN   55355
   (612) 693-2887
   Director

Mellette County
Planning Department
County Courthouse
White River, SD   57579
   (605) 259-3291
   Director

Mendocino County
Planning Department
County Courthouse
Ukiah, CA   95482
   (707) 468-4011
   Director

Menominee County
Planning Department
County Courthouse
Keshena, WI   54135
   (715) 799-3311
   Director

Merced County Association of Govern-
ments
P.O. Box 2201
Merced, CA   95340
   (209) 723-3153
   Director

Merced County
Planning Department
2222 M Street
Merced, CA   95340
   (209) 726-7254
   Chief: Michael Pellock

Meriden City

Planning Department
City Hall
Room 30
Meriden, CT 06450
   (203) 634-0003
   EIS Contact: Dominick Caruso

Merrimack Valley Planning Commission
5 Washington Street
Haverhill, MA 01830
   (617) 374-0519
   Project Manager: Jeffrey Riotte

Metropolitan Area Planning Council
44 School Street
Boston, MA 02108
   (617) 523-2454
   Director

Metropolitan Association of Urban
   Designers and Environmental
   Planners
P.O. Box 722
Church Street Station
New York, NY 10008
   (212) 227-6970
   President: Carl Berkowitz

Metropolitan Council
300 Metro Square Building
St. Paul, MN 55101
   (612) 291-6452
   Director

Metropolitan Inter-County Council
55 Sherburne Avenue
St. Paul, MN 55103
   (612) 222-1749
   Director

Miami County
Planning Department
County Courthouse
Peru, IN 46970
   (317) 472-3994
   Director

Michiana Area Council of Governments
County City Building
South Bend, IN 46601
   (219) 287-1829
   Director

Michigan Association of Counties
319 West Lenawea Street
Lansing, MI 48933
   (517) 372-5374
   Executive Director: Barry McGuire

Michigan
Department of Agriculture
Soil and Water Management Division
Fifth Floor
Lewis Cass Building
P.O. Box 30017
Lansing, MI 48909
   (517) 373-1050
   Chief: Stanley Quackenbush

Michigan
Department of Natural Resources
Box 30028
Lansing, MI 48909
   (517) 373-1220
   Director: Howard Tanner

Michigan
Department of Natural Resources
Environmental Enforcement Division
P.O. Box 30028
Lansing, MI 48909
   (517) 373-3503
   Branch Chief: Dr. Donald Inman

Michigan
Department of Natural Resources
Environmental Protection
P.O. Box 30028
Lansing, MI 48909
   (517) 373-1220
   Bureau Chief: William Turney

Michigan
Department of Natural Resources
Environmental Services Division
P.O. Box 30028
Lansing, MI 48909
   (517) 373-1220
   Chief: Gary Guenther

Michigan Municipal League
P.O. Box 1487
Ann Arbor, MI 48106
   (313) 662-3246
   Director: John Patriarche

Michigan United Conservation Clubs
Box 30235
Lansing, MI 48909
  (517) 371-1041
  President: Glenn Corbett

Middle Peninsula Planning District
  Commission
P.O. Box 286
Saluda, VA 23149
  (804) 758-2312
  Director

Middlesex County
Planning Department
County Courthouse
Cambridge, MA 01742
  (617) 494-4000
  Director

Middlesex County
Planning Department
County Courthouse
Saluda, VA 23149
  (804) 758-5311
  Director

Mid-East Commission
P.O. Box 1218
310 West Main Street
Washington, NC 27889
  (919) 946-8043
  Executive Director: Daneel le Roux

Midstate Regional Planning Agency
P.O. Box 139
Middletown, CT 06457
  (203) 347-7214
  Director

Milford City
Planning Department
One Polizzi Plaza
Milford, CT 06460
  (203) 878-1731
  City Planner: Wade Pierce

Mille Lacs County
Planning Department
County Courthouse
Milaca, MN 56353
  (612) 983-6282

  Zoning Officer: Mahlon Dahl

Milwaukee City
Planning Department
P.O. Box 324
Milwaukee, WI 53202
  (414) 278-3200
  EIS Contact: Dr. Rodolfo Salcedo

Milwaukee County
Department of Public Works
Environmental Services
Courthouse Annex
907 North 10th Street
Milwaukee, WI 53233
  (414) 278-4874
  Director: Fred Rehm

Mineral County
Planning Department
P.O. Box 517
Superior, MT 59872
  (406) 822-4541
  Director

Miner County
Planning Department
County Courthouse
Howard, SD 57349
  (605) 772-4671
  Director

Minneapolis City
Planning Department
Room 210
City Hall
Minneapolis, MN 55415
  (612) 348-2597
  EIS Contact: Michael Cronin

Minnehaha County
Planning Department
County Courthouse
Sioux Falls, SD 57102
  (605) 336-2350
  Director

Minnesota Association of Counties
2305 Ford Parkway
St. Paul, MN 55116
  (612) 698-4212
  Executive Director: James Shipman

Minnesota Association of Soil and
  Water Conservation Districts
Fairmount, MN 58030
  (218) 479-2759
  President: Robert Wetherbee

Minnesota Conservation Federation
Room 218C
790 South Cleveland Avenue
St. Paul, MN 55116
  (612) 690-3077
  Director

Minnesota
Department of Natural Resources
300 Centennial Building
658 Cedar Street
St. Paul, MN 55155
  (612) 296-2549
  Commissioner: Joseph Alexander

Minnesota
Department of Natural Resources
Bureau of Land
658 Cedar Street
St. Paul, MN 55155
  (612) 296-0660
  Administrator: Richard Hultengren

Minnesota
Department of Natural Resources
State Soil and Water Conservation
  Board
Box 19
Centennial Building
St. Paul, MN 55155
  (612) 296-3767
  Executive Director: Vernon Renert

Minnesota League of Cities
300 Hanover Building
480 Cedar Street
St. Paul, MN 55101
  (612) 222-2861
  Executive Director: Donald Slater

Minnesota
Pollution Control Agency
1935 West County Road B2
Roseville, MN 55113
  (612) 296-7200
  EIS Contact: Janet Cain

Minnesota
Pollution Control Agency
Brainerd Regional Office
615 Oak Street
Brainerd, MN 56401
  (218) 828-2492
  Director: Larry Shaw

Minnesota
Pollution Control Agency
Detroit Lakes Regional Office
116 East Front Street
Detroit Lakes, MN 56501
  (218) 847-2165
  Director: Willis Mattison

Minnesota
Pollution Control Agency
Duluth Regional Office
1015 Torrey Building
Duluth, MN 55802
  (218) 723-4660
  Director: John Pegors

Minnesota
Pollution Control Agency
Marshall Regional Office
P.O. Box 286
S.W. University
Marshall, MN 56258
  (507) 537-7416
  Director: Lawrence Johnson

Minnesota
Pollution Control Agency
Rochester Regional Office
821 Third Avenue, S.E.
Rochester, MN 55901
  (507) 285-7343
  Director: Larry Landherr

Minnesota Valley Council of Governments
202 East Jackson Street
Box 328
Mankato, MN 56001
  (507) 625-3161
  Director

Mississippi
Air and Water Pollution Control
  Commission

P.O. Box 827
Jackson, MS 39205
    (601) 354-2550
    Executive Director: Charles Chisolm

Mississippi Association of Conser-
    vation Districts
6210 Hanging Moss Road
Jackson, MS 39216
    (601) 366-3932
    President: Bowman Virden

Mississippi Association of
    Supervisors
P.O. Box 1314
Jackson, MS 39205
    (601) 353-2741
    Presidential Assistant: A.J. Foster

Mississippi Municipal Association
230 Barefield Complex
Jackson, MS 39202
    (601) 353-5854
    Executive Director: Patrick Dunne

Mississippi River Regional Planning
    Commission
315 South Front Street
Grandview Building
La Crosse, WI 54601
    (608) 784-5516
    EIS Contact: Doug Venable

Mississippi
State Soil and Water Conservation
    Commission
754 North President Street
Jackson, MS 39201
    (601) 354-7469
    Executive Director: Gale Martin

Missoula County
Planning Office
301 West Alder
Missoula, MT 59801
    (406) 721-5700
    Director: William Walton

Missoula Planning Board
301 West Alder
Missoula, MT 59801
    (406) 728-1561

Director

Missouri Association of Counties
P.O. Box 234
Jefferson City, MO 65101
    (314) 634-2120
    Executive Director: Tony Hiesberger

Missouri Association of Soil and Water
    Conservation Districts
Naylor, MO 63953
    (314) 399-2450
    President: Fred Moutrie

Missouri
Department of Conservation
P.O. Box 180
Jefferson City, MO 65101
    (314) 751-4115
    Director: Larry Gale

Missouri
Department of Conservation
Environmental Services
P.O. Box 180
Jefferson City, MO 65102
    (314) 751-4115
    Supervisor: William Dieffenbach

Missouri
Division of Environmental Quality
P.O. Box 1368
Jefferson City, MO 65101
    (314) 751-3241
    Director: James Odendahl

Missouri Municipal League
1913 William Street
Jefferson City, MO 65101
    (314) 635-9134
    Executive Director: Jay Bell

Mitchell County
Planning Department
County Courthouse
Bakersville, NC 27807
    (704) 688-2161
    Director

Modesto City
Planning Department
P.O. Box 642

City Hall
11th and H Streets
Modesto, CA 95353
  (209) 577-5273
  EIS Contact: George Osner

Modoc County
Planning Department
County Courthouse
Alturas, CA 96101
  (916) 233-2215
  Director

Mohawk Valley Economic Development
  District
26 West Main Street
P.O. Box 69
Mohawk, NY 13407
  (315) 866-4671
  Director

Mono County
Planning Department
County Courthouse
Bridgeport, CA 93517
  (714) 932-7911
  Director

Monroe County
Planning Department
County Courthouse
Bloomington, IN 47401
  (812) 336-3424
  Director

Monroe County
Planning Department
301 County Office Building
Rochester, NY 14614
  (716) 428-5469
  Associate Planner: W.A. Frazier, Jr.

Monroe County
Planning Department
County Courthouse
Sparta, WI 54656
  (608) 269-4411
  Director

Montachusett Regional Planning
  Commission
150 Main Street

Fitchburg, MA 01420
  (617) 345-2216
  EIS Contact: Laila Michaud

Montana Association of Conservation
  Districts
7 Edwards
Helena, MT 59601
  (406) 443-5711
  President: Walt Dion

Montana Association of Counties
1802-11th Avenue
Helena, MT 59601
  (406) 442-5209
  Executive Director: Dean Zinnecker

Montana
Department of Health and Environmental
  Sciences
Environmental Sciences Division
Cogswell Building
Helena, MT 59601
  (406) 587-3946
  Administrator: Benjamin Wake

Montana
Department of Natural Resources and
  Conservation
32 South Ewing
Helena, MT 59601
  (406) 449-3712
  Director: Ted Doney

Montana
Environmental Quality Council
State Capitol
Helena, MT 59601
  (406) 449-3742
  Executive Director: Terrence
    Carmody

Montana League of Cities and Towns
P.O. Box 1704
Helena, MT 59601
  (406) 442-8768
  Executive Director: Dan Mizner

Monterey County
Planning Department
P.O. Box 1204
Salinas, CA 93902

(408) 424-8611
Director: E.W. DeMars

Montgomery County
Planning Department
County Courthouse
Crawfordsville, IN 47933
  (317) 362-6302
  Director

Montgomery County
Planning Department
County Courthouse
Rockville, MD 20850
  (301) 279-1284
  Director

Montgomery County
Planning Department
County Annex Building
Fonda, NY 12068
  (518) 853-3431
  EIS Contact: Jack Jowett

Montgomery County
Planning Department
County Courthouse
Troy, NC 27371
  (919) 576-4211
  Director

Montgomery County
Planning Department
P.O. Box 806
Christianburg, VA 24073
  (703) 382-2644
  Director

Moody County
Planning Department
County Courthouse
Flandreau, SD 57028
  (605) 997-2469
  Director

Moore County
Planning Department
County Courthouse
Carthage, NC 28327
  (919) 947-2396
  Director

Morgan County
Planning Department
County Courthouse
Martinsville, IN 46151
  (317) 342-7124
  Director

Morrison County
Zoning and Planning Office
Old Courthouse Building
Little Falls, MN 56345
  (612) 632-9215
  EIS Contact: Kathy Kendall

Mountain View City
Planning Department
P.O. Box 10
Mountain View, CA 94042
  (415) 967-7211
  EIS Contact: Steve Unangst

Mount Rogers Planning District
 Commission
1021 Terrace Drive
Marion, VA 24354
  (703) 783-5103
  Director

Mount Vernon City
Planning Department
City Hall
Mount Vernon, NY 10550
  (914) 668-2200
  Director

Mower County
Planning Department
County Courthouse
Austin, MN 55912
  (507) 433-2077
  Director

Muncie City
Planning Department
City Hall
Muncie, IN 47302
  (317) 747-4831
  Director

Murray County
Planning Department

County Courthouse
Slayton, MN 56172
  (507) 836-6158
  Director

Musselshell County
Planning Department
County Courthouse
Roundup, MT 59072
  (406) 323-1104
  Director

Nantucket County
Planning Department
County Courthouse
Nantucket, MA 02554
  (617) 228-0925
  Director

Nantucket Planning and Economic
  Development Commission
Broad Street
Nantucket, MA 02554
  (617) 228-9625
  Director: William Klein

Napa County
Planning Department
1115 First Street
Napa, CA 94558
  (707) 253-4421
  Director

Nash County
Planning Department
P.O. Drawer G
Nashville, NC 27856
  (919) 459-3028
  Director: Robert Bridwell

Nassau County
Planning Commission
222 Willis Avenue
Mineola, NY 11501
  (516) 535-2244
  EIS Contact: Robert Berry

Nassau-Suffolk Regional Planning
  Board
Lee Dennison Building
Veterans Memorial Highway
Hauppauge, NY 11787

  (516) 724-1919
  Director

National Association of Conservation
  Districts
1025 Vermont Avenue, N.W.
Washington, DC 20005
  (202) 347-5995
  President: Lyle Bauer

National Association of Counties
  Research Foundation
1735 New York Avenue, N.W.
Washington, DC 20006
  (202) 785-9577
  President: Charlotte Williams

National Council for Environmental
  Balance
4169 Westport Road
P.O. Box 7732
Louisville, KY 40207
  (502) 896-8731
  President: Dr. Irwin Tucker

National Council of the Paper Industry
  for Air and Stream Improvement
260 Madison Avenue
New York, NY 10016
  (212) 883-8083
  Executive Vice President: Dr.
    Isaiah Gellman

National Farmers Union
Box 39251
12025 East 45th Avenue
Denver, CO 80251
  (303) 371-1760
  President: Tony Dechant

National Geographic Society
17th and M Streets, N.W.
Washington, DC 20036
  (202) 857-7000
  President: Robert Doyle

National Grange
1616 H Street, N.W.
Washington, DC 20006
  (202) 628-3507
  Master: John Scott

National Wildlife Federation
1412-16th Street, N.W.
Washington, DC 20036
   (202) 797-6800
   President: Dr. F.R. Scroggin

Natural Resources Council of America
Box 20
Tracy Landing, MD 20869
   (301) 261-5277
   Chairman: Brock Evans

Natural Resources Council of Maine
51 Chapel Street
Augusta, ME 04330
   (207) 622-3101
   President: Peter Heimann

Natural Resources Defense Council
122 East 42nd Street
New York, NY 10017
   (212) 949-0049
   Executive Director: John Adams

Nature Conservancy
Suite 800
1800 North Kent Street
Arlington, VA 22209
   (703) 841-5300
   President: Patrick Noonan

Nebraska Association of County
   Officials
103 Executive Building
521 South 14th Street
Lincoln, NE 68508
   (402) 474-3328
   Executive Director: Jack Mills

Nebraska Association of Resources
   Districts
Suite 308
Sharp Building
Lincoln, NE 68508
   (402) 432-8506
   President: Leon Halstead

Nebraska
Department of Environmental Control
State House Station
Box 94877
Lincoln, NE 68509

   (402) 471-2186
   Director: Dan Drain

Nebraska League of Municipalities
1335 L Street
Lincoln, NE 68508
   (402) 432-2829
   Director: David Chambers

Nebraska Natural Resources Commission
301 Centennial Mall South
P.O. Box 94876
Lincoln, NE 68509
   (402) 471-2081
   Executive Secretary: Dayle
      Williamson

Nelson County
Planning Department
County Courthouse
Lovingston, VA 22949
   (804) 263-4873
   Director

Neuse River Council of Governments
1404 Neuse Boulevard
P.O. Box 1717
New Bern, NC 28560
   (919) 638-3185
   Director

Nevada Association of Conservation
   Districts
Jiggs, NV 89827
   (702) 753-6338
   President: Fred Zaga

Nevada Association of County Commissioners
Clark County Courthouse
200 East Carson Street
Las Vegas, NV 89011
   (702) 588-2463
   Executive Secretary: Thalia Dondero

Nevada County
Planning Department
HEW Building
10433 Willow Valley Road
Nevada City, CA 95959
   (916) 265-2461
   Principal Planner: Tom Parilo

Nevada
Department of Conservation and
   Natural Resources
Capitol Complex
Nye Building
201 South Fall Street
Carson City, NV 89710
   (702) 885-4360
   Director: Norman Hall

Nevada
Department of Conservation and
   Natural Resources
Division of Conservation Districts
Capitol Complex
Nye Building
201 South Fall Street
Carson City, NV 89710
   (702) 885-5414
   Administrative Officer: Ted Bendure

Nevada
Department of Conservation and
   Natural Resources
Division of Environmental Protection
Capitol Complex
Nye Building
201 South Fall Street
Carson City, NV 89710
   (702) 885-4670
   Administrator: Ernie Gregory

Nevada
Department of Conservation and
   Natural Resources
State Environmental Commission
Capitol Complex
Nye Building
201 South Fall Street
Carson City, NV 89710
   (702) 885-4670
   Executive Secretary: Ken Boyer

Nevada League of Cities
P.O. Box 2307
Carson City, NV 89701
   (702) 882-2121
   Executive Secretary: Gentty
      Etcheverry

New Bedford City
Planning Department

City Hall
New Bedford, MA 02740
   (617) 999-2931
   EIS Contact: Robert Bowcock

New Britain City
Planning Department
City Hall
New Britain, CT 06051
   (203) 224-2491
   Director

New England River Basins Commission
53 State Street
First Floor
Boston, MA 02109
   (617) 223-6244
   Chairman: John Ehrenfeld

New Hampshire Association of Conservation Commissions
5 South State Street
Concord, NH 03301
   (603) 224-9945
   President: Joan McGoldrick

New Hampshire Association of Conservation Districts
Loudon, NH 03301
   (603) 267-8050
   President: Robert Hibbard

New Hampshire Association of Counties
163 North Main Street
Concord, NH 03301
   (603) 228-0331
   Executive Secretary: Peter Spaulding

New Hampshire Municipal Association
P.O. Box 617
Concord, NH 03301
   (603) 224-7447
   Executive Director: John Andrews

New Hanover County
Planning Department
County Courthouse
Wilmington, NC 28401
   (919) 763-3688
   Director

New Haven City

Planning Commission
157 Church Street
New Haven, CT   06510
   (203) 562-0151
   Executive Director: John McGuerty

New Jersey Association of Counties
Suite 3-B
120 Sanhican Drive
Trenton, NJ   08618
   (609) 394-3467
   Executive Director: Jack Lamping

New Jersey Conservation Foundation
300 Mendham Road
Morristown, NJ   07960
   (201) 539-7540
   President: Gordon Millspaugh, Jr.

New Jersey
Department of Agriculture
Soil and Water Conservation
P.O. Box 1888
Trenton, NJ   08625
   (609) 292-3976
   Coordinator: Samuel Race

New Jersey
Department of Environmental Pro-
   tection
P.O. Box 1390
Trenton, NJ   08625
   (609) 292-2885
   Commissioner: Jerry English

New Jersey
Department of Environmental Pro-
   tection
Division of Environmental Quality
P.O. Box CN027
Trenton, NJ   08625
   (609) 292-5383
   Director: George Tyler

New Jersey
Department of Environmental Pro-
   tection
Office of Environmental Analysis
P.O. Box 1390
Trenton, NJ   08625
   (609) 292-2938
   Chief: Roland Yunghams

New Jersey State Federation of Sports-
   men's Clubs
Box 488
Freehold, NJ   07728
   (201) 536-3730
   President: John Volk

New Jersey State League of Municipalities
433 Bellevue Avenue
Room D 403
Trenton, NJ   08618
   (609) 695-3481
   Executive Director: Robert Fust

New Kent County
Planning Department
P.O. Box 50
New Kent, VA   23124
   (804) 966-9861
   County Administrator: Royal Wood

New Mexico Association of Counties
P.O. Box 1748
Sante Fe, NM   87501
   (505) 983-2101
   Executive Director: Philip
      Larragoite

New Mexico Association of Natural
   Resources Conservation Districts
Stanley, NM   87056
   (505) 832-4892
   President: David King

New Mexico Conservation Coordinating
   Council
P.O. Box 142
Alburquerque, NM   87103
   President: A. Cowan Collins

New Mexico
Environmental Improvement Division
P.O. Box 968
Sante Fe, NM   87503
   (505) 827-5271
   Director: Thomas Baca

New Mexico
Environmental Improvement Division
Environmental Review
P.O. Box 968

Sante Fe, NM 87503
  (505) 827-5271
  Coordinator: G. Carl Selnick

New Mexico Municipal League
P.O. Box 846
Sante Fe, NM 87501
  (505) 982-5573
  Executive Director: William
    Fulginiti

New Mexico Wilderness Study
  Committee
P.O. Box 801
Silver City, NM 88061
  (505) 388-4326
  Chairman: Bob Langsenkamp

Newport News City
Planning Department
City Hall
Newport News, VA 23601
  (804) 247-8411
  Director

New River Valley Planning District
  Commission
1612 Wadsworth Street
P.O. Box 3726
Radford, VA 24141
  (703) 639-9313
  Regional Planner: Robert West

New Rochelle City
Planning Department
City Hall
New Rochelle, NY 10801
  (914) 632-2021
  Director

Newton City
Community Development Program
1000 Commonwealth Avenue
Newton, MA 02159
  (617) 552-7135
  Environmental Review Officer:
    Dale Silin

Newton County
Planning Department
County Courthouse
Kentland, IN 47951

  (219) 474-5842
  Director

New York
Bureau of Environmental Protection
Department of Law
State of New York
Two World Trade Center
New York, NY 10047
  (212) 488-5123
  Assistant Attorney General: Philip
    Weinberg

New York City
Planning Department
80 Centre Street
New York City, NY 10007
  (212) 566-4446
  Director

New York County
Planning Department
County Courthouse
New York, NY 10001
  (212) 374-8361
  Director

New York
Department of Environmental Conservation
50 Wolf Road
Albany, NY 12233
  (518) 457-3446
  Commissioner: Peter Berle

New York
Department of Environmental Conser-
  vation
Region 1
Building 40
State University of New York
Stony Brook, NY 11794
  (516) 751-7900
  Director: Donald Middleton

New York
Department of Environmental Conservation
Region 2
2 World Trade Center
New York, NY 10047
  (212) 488-2577

Director: Terry Agriss

New York
Department of Environmental Conser-
    vation
Region 3
21 South Putt Corners Road
New Paltz, NY 12561
    (914) 255-5453
    Director: Paul Keller

New York
Department of Environmental Conser-
    vation
Region 4
50 Wolf Road
Albany, NY 12233
    (518) 457-5861
    Director: David Perriman

New York
Department of Environmental Conser-
    vation
Region 5
Ray Brook, NY 12977
    (518) 891-1370
    Director: Thomas Monroe

New York
Department of Environmental Conser-
    vation
Region 6
317 Washington Street
St. Watertown, NY 13601
    (315) 782-0100
    Director: John Wilson

New York
Department of Environmental Conser-
    vation
Region 7
7481 Henry Clay Boulevard
Liverpool, NY 13088
    (315) 473-8301
    Director: William Hicks

New York
Department of Environmental Conser-
    vation
Region 8
P.O. Box 57
Avon, NY 14414

    (716) 226-2466
    Director: Erie Seiffer

New York
Department of Environmental Conser-
    vation
Region 9
584 Delaware Avenue
Buffalo, NY 14202
    (716) 842-5824
    Director: William Friedman

New York State Association of Conser-
    vation Commissions
12 Main Street
Hamburg, NY 14075
    (716) 649-0950
    President: Julie Krug

New York Association of Counties
150 State Street
Albany, NY 12207
    (518) ·465-1473
    Executive Director: Edwin Crawford

New York State Conference of Mayors
6 Elk Street
Albany, NY 12207
    (518) 463-1185
    Executive Director: Gordon Perry

New York State Conservation Council
8 East Main Street
Ilion, NY 13357
    (315) 894-3302
    President: Robert Young

New York
State Department of Public Services
Office of Environmental Planning
Empire State Plaza
Building 3
Albany, NY 12223
    (518) 474-1677
    Director: Robert Vessels

New York State Soil and Water Conser-
    vation Committee
142 Emerson Hall
Cornell University
Ithaca, NY 14853
    (607) 256-4420

Executive Secretary: Willard Croney

Niagara County
Planning Department
County Courthouse
Lockport, NY 14094
(716) 434-0616
Director

Niagara Falls City
Planning Department
City Hall
Niagara Falls, NY 14302
(716) 278-8000
Director

Nicollet County
Office of Planning and Zoning
County Courthouse
St. Peter, MN 56082
(507) 931-6800
Zoning Administrator: Phil Lutzi

Noble County
Planning Department
County Courthouse
Albion, IN 46701
(219) 636-2658
Director

Nobles County
Planning Department
County Courthouse
Worthington, MN 56187
(507) 376-4151
Director

Norfolk City
Planning Department
1101 City Hall Boulevard
Norfolk, VA 23501
(804) 441-2471
Director

Norfolk County
Planning Department
County Courthouse
Dedham, MA 02026
(617) 326-1600
Director

Norman County

Planning Department
County Courthouse
Ada, MN 56510
(218) 784-7131
Director

Northampton County
Planning Department
County Courthouse
Jackson, NC 27845
(919) 534-2221
Director

Northampton County
Planning Department
County Courthouse
Eastville, VA 23347
(804) 678-5148
Director

North Carolina Association of County
Commissioners
P.O. Box 1488
Raleigh, NC 27602
(919) 832-2893
Executive Director: Ronald Aycock

North Carolina Association of Soil and
Conservation Districts
Route 3
Fuquay-Varina, NC 27256
(919) 552-5038
President: Steward Adcock

North Carolina
Department of Natural Resources and
Community Development
P.O. Box 27687
Raleigh, NC 27611
(919) 733-4984
Environmental Manager: R.F. McRorie

North Carolina
Department of Natural Resources and
Community Development
Soil and Water Conservation Commission
512 North Salisbury Street
Raleigh, NC 27611
(919) 733-2302
Chief: Grady Lane

North Carolina League of Municipalities

P.O. Box 3069
Raleigh, NC 27602
(919) 834-1311
Executive Director: Leigh Wilson

North Central Wisconsin Regional
  Planning Commission
2100 Main Street
Stevens Point, WI 54481
(715) 346-3311
Director

Northcoast Environmental Center
1091 H Street
Arcata, CA 95521
(707) 822-6918
Coordinator: Tim McKay

North Dakota
Association of Counties
P.O. Box 417
Bismark, ND 58501
(701) 258-4481
Executive Secretary: Ron Soderberg

North Dakota
Department of Health
Environmental Control
Bismark, ND 58505
(701) 224-2371
Chief: Gene Christianson

North Dakota League of Cities
P.O. Box 2235
Bismark, ND 58501
(701) 223-3518
Executive Director: Arne Boyum

North Dakota Natural Science Society
P.O. Box 1672
Jamestown, ND 58401
(701) 252-5363
President: Dr. John Bluemle

Northeast Conservation Law Enforce-
  ment Chiefs' Association
Law Enforcement Unit
Connecticut Department of Environ-
  mental Protection
Room 247
State Office Building
Hartford, CT 16115

(203) 566-3978
President: Frederick Pogmore

Northeastern Connecticut Regional
  Planning Agency
P.O. Box 198
Brooklyn, CT 06234
(203) 774-1253
Executive Director: Gerald McCarthey

Northeastern Indiana Regional Coordi-
  nating Council
City County Building
One Main Street
Fort Wayne, IN 46802
(219) 423-7309
Director

Northern Environment Committee
Executive Committee
Box 39
Route 1
Minong, WI 54859
(715) 466-2480
Chairman: Charles Stoddard

Northern Middlesex Area Commission
144 Merrimack Street
Lowell, MA 01852
(617) 454-8021
Director

Northern Neck Planning District
  Commission
Drawer H
Callao, VA 22435
(804) 529-7400
Director

Northern Rockies Action Group
9 Placer Street
Helena, MT 59601
(406) 442-6615
Coordinator: Bill Bryan

Northern Virginia Conservation
  Council
P.O. Box 304
Annandale, VA 22003
(703) 941-5321
President: Thomas Gause

Northern Virginia Planning District
  Commission
7309 Arlington Boulevard
Falls Church, VA 22042
  (703) 573-2210
  EIS Contact: Martha Mason Semmes

Northumberland County
Planning Department
County Courthouse
Heathsville, VA 22473
  (804) 580-7666
  Director

Northwestern Connecticut Regional
  Planning Agency
Sackett Hill Road
P.O. Box 30
Warren, CT 06754
  (203) 868-7341
  EIS Contact: Charles Boster

Northwestern Indiana Regional
  Planning Commission
8149 Kennedy Avenue
Highland, IN 46322
  (219) 923-1060
  Director

Northwest Regional Development
  Commission
425 Woodland Avenue
Crookston, MN 56716
  (218) 281-1396
  Director

Northwest Regional Planning
  Commission
302 Walnut Street
Spooner, WI 54801
  (715) 635-2197
  EIS Contact: Philip Scherer

Northwest Wildlife Law Enforcement
  Association
U.S. Fish and Wildlife Service
500 N.E. Multnomah Street
Portland, OR 97232
  (503) 231-6125
  President: Lawrence Wills

Norwalk City

Planning Department
12700 Norwalk Boulevard
Norwalk, CA 90650
  (213) 868-3254
  Senior Planner: Art Rangel

Norwalk City
Planning Department
City Hall
Norwalk, CT 06854
  (203) 838-7531
  Director

Nottoway County
Planning Department
County Courthouse
Nottoway, VA 23955
  (804) 645-8696
  Director

Oakland City
Planning Department
City Hall
Oakland, CA 94612
  (415) 273-9000
  Director

Oak Ridge National Laboratory
Ecosystems Analysis Data Center
P.O. Box X
Oak Ridge, TN 37830
  (615) 483-8611
  Julia Watts

Oak Ridge National Laboratory
Energy and Environmental Response
  Center
P.O. Box X
Oak Ridge, TN 37830
  (615) 483-8611
  Director: D.J. Wilkes

Oak Ridge National Laboratory
Environmental Impact Section
P.O. Box X
Oak Ridge, TN 37830
  (615) 483-8611
  Section Head: T.H. Row

Oak Ridge National Laboratory
Environmental Sciences Division
P.O. Box X

Building 1505
Oak Ridge, TN 37830
  (615) 483-8611
  Richard Olson

Oak Ridge National Laboratory
Nuclear Safety Information Center
P.O. Box Y
Oak Ridge, TN 37830
  (615) 483-8611
  Joel Buchanan

Oak Ridge National Laboratory
Regional and Urban Studies Infor-
  mation Center
P.O. Box X
Oak Ridge, TN 37830
  (615) 483-8611
  Director: A.S. Lobel

Oconto County
Planning Department
County Courthouse
Oconto, WI 54153
  (414) 834-5322
  Director

Ohio County
Planning Department
County Courthouse
Rising Sun, IN 47040
  (812) 438-2610
  Director

Ohio County Commissioners' Assoc-
  iation
41 South High Street
Neil House
M-58
Columbus, OH 43215
  (614) 221-5627
  Executive Director: A.R. Masler

Ohio Environmental Council
850 Michigan Avenue
Columbus, OH 43215
  (614) 221-0898
  President: Dr. Robert Alrutz

Ohio
Environmental Protection Agency
Box 1049

361 East Broad Street
Columbus, OH 43216
  (614) 466-7232
  EIS Coordinator: Beth Whitman

Ohio
Environmental Protection Agency
Office of Land Pollution Control
Box 1049
361 East Broad Street
Columbus, OH 43216
  (614) 466-8307
  Jim Michael

Ohio Federation of Soil and Water
  Conservation Districts
RT 3
Versailles, OH 45380
  (419) 336-7442
  President: Arthur Brandt

Ohio Municipal League
Suite 105
41 South High Street
Columbus, OH 43215
  (614) 221-4349
  Executive Director: John Coleman

Okanogan County
Planning Department
P.O. Box 1009
Okanogan, WA 98840
  (509) 422-3521
  EIS Contact: Steve Burger

Okanogan County Regional Planning
  Commission
Box 1009
Okanogan, WA 98840
  (509) 422-3301
  Director

Oklahoma Association of Conservation
  Districts
Box 579
Pauls Valley, OK 73075
  (405) 238-3393
  President: Jack Grimmett

Oklahoma County Commissioners Association
c/o Washita County Commissioners

Cordell, OK 73632
(405) 832-2284
Secretary: Harvey Weichel

Oklahoma
Department of Pollution Control
Box 53504
N.E. 10th and Stonewall
Oklahoma City, OK 73117
(405) 271-4677
Director: Lawrence Edminson

Oklahoma
Department of Wildlife Conservation
Environmental Services
1801 North Lincoln
P.O. Box 53465
Oklahoma City, OK 73152
(405) 521-3851
Chief: Ricardo Gomez

Oklahoma Municipal League
201 N.E. 23rd Street
Oklahoma City, OK 73105
(405) 528-7515
Executive Director: Donald Rider

Old Colony Planning Council
232 Main Street
Brockton, MA 02401
(617) 583-1833
Director

Olmsted County
Planning Department
1421 S.E. 3rd Avenue
Rochester, MN 55901
(507) 285-8232
EIS Contact: Gary Lueders

Oneida County
Planning Department
County Courthouse
Utica, NY 13503
(315) 798-5800
Director

Oneida County
Planning Department
County Courthouse
P.O. Box 400
Rhinelander, WI 54501

(715) 369-2727
EIS Contact: John Vanney

Onondaga County
Environmental Management Board
1100 Civic Center
421 Montgomery Street
Syracuse, NY 13202
(315) 425-2640
EIS Contact: Robert Deyle

Onsiow County
Planning Department
107 New Bridge Street
Jacksonville, NC 28540
(919) 347-4717
Director

Ontario City
Planning Department
City Hall
Ontario, CA 91761
(714) 986-1151
Director

Ontario County
Planning Department
County Courthouse
Canandaigua, NY 14424
(315) 394-7070
Director

Open Lands Project
53 West Jackson Boulevard
Chicago, IL 60604
(312) 427-4256
Executive Director: Judith Stockdale

Orange City
Planning Department
City Hall
Orange, CA 92666
(714) 532-0321
Director

Orange County
Planning Department
County Courthouse
Santa Ana, CA 92701
(714) 834-2200
Director

Orange County
Planning Department
County Courthouse
Paoli, IN 47454
  (812) 723-2649
  Director

Orange County
Planning Department
124 Main Street
1887 Building
Goshen, NY 10924
  (914) 294-5151
  EIS Contact: Richard Jones

Orange County
Planning Department
County Courthouse
Hillsborough, NC 27278
  (919) 732-8181
  Director

Orange County
Planning Department
County Courthouse
Orange, VA 22960
  (703) 672-3313
  Director

Oregon Association of Counties
P.O. Box 12729
Salem, OR 97309
  (503) 585-8351
  Executive Director: Jerry Orrick

Oregon Association of Soil and Water
  Conservation Districts
2540 Olson Road
Tillamook, OR 97141
  (503) 842-2874
  President: Ernest Josi

Oregon
Department of Environmental Quality
522 S.W. Fifth Avenue
P.O. Box 1760
Portland, OR 97207
  (503) 229-5395
  Director: William Young

Oregon
Department of Environmental Quality

522 S.W. Fifth Avenue
P.O. Box 1760
Portland, OR 97207
  (503) 229-6403
  Intergovernmental Coordinator:
    Robert Jackman

Oregon Environmental Council
Office of the Director
2637 S.W. Water Avenue
Portland, OR 97201
  (503) 222-1963
  Executive Director: John Platt

Oregon League of Cities
P.O. Box 928
Salem, OR 97308
  (503) 588-6466
  Executive Director: Stephen Bauer

Oregon Wilderness Coalition
P.O. Box 3066
Eugene, OR 97403
  (503) 686-5014
  President: Holway Jones

Orleans County
Planning Department
151 Platt Street
Albion, NY 14411
  (716) 589-7065
  Director: Patrick Rountree

Oshkosh City
Planning Department
City Hall
Oshkosh, WI 54901
  (414) 424-0274
  Director

Oswego County
Planning Department
County Building
46 East Bridge Street
Oswego, NY 13126
  (315) 349-3385
  Director

Otsego County
Planning Department
County Courthouse
Cooperstown, NY 13326

(607) 547-4276
Director

Otter Tail County
Land and Resources Management
County Courthouse
Fergus Falls, MN 56537
    (218) 739-2271
    EIS Contact: Malcolm Lee

Ouabache Regional Development
  Commission
23 South Broadway
Peru, IN 46970
    (317) 472-3936
    Director

Outagamie County
Planning Department
County Courthouse
Appleton, WI 54911
    (414) 739-7673
    Director

Owen County
Planning Department
County Courthouse
Spencer, IN 47460
    (812) 829-2325
    Director

Oxnard City
Planning Department
305 West Third Street
Oxnard, CA 93030
    (805) 486-2601
    Director

Ozaukee County
Planning Department
County Courthouse
Port Washington, WI 53074
    (414) 284-9411
    Director

Pacific County
Planning Department
County Courthouse
Box 66
South Bend, WA 98586
    (206) 875-5591
    EIS Contact: Ken Kimura

Pacific County Regional Planning
  Council
P.O. Box 66
South Bend, WA 98586
    (206) 875-5591
    Director

Page County
Planning Department
County Courthouse
Luray, VA 22835
    (703) 743-4064
    Director

Palo Alto City
Planning Department
City Hall
Palo Alto, CA 94301
    (415) 329-2149
    EIS Contact: Robert Brown

Pamilico County
Planning Department
P.O. Box 186
Bayboro, NC 28515
    (919) 745-3861
    EIS Contact: Gene Broughton

Park County
Planning Department
County Courthouse
Livingston, MT 59047
    (406) 222-0450
    Director

Parke County
Planning Department
County Courthouse
Rockville, IN 47872
    (317) 569-5132
    Director

Pasadena City
Planning Department
City Hall
Pasadena, CA 91109
    (213) 577-4000
    Director

Pasquotank County
Planning Department
P.O. Box 272

Elizabeth City, NC 27909
  (919) 338-0175
  Director

Patoka Lake Regional Planning
  Commission
Courthouse
Third Floor
P.O. Box 690
Jasper, IN 47546
  (812) 482-4645
  Director

Patrick County
Planning Department
P.O. Box 466
Stuart, VA 24171
  (703) 694-6094
  Director

Pee Dee Council of Governments
227 North Main Street
P.O. Box 728
Troy, NC 27371
  (919) 576-6261
  Director

Pender County
Planning Department
P.O. Box 832
Burgaw, NC 28425
  (919) 259-5461
  Director: Alvin Midgette

Pend Oreille County
Planning Department
County Courthouse
Newport, WA 99156
  (509) 447-4119
  Director

Peninsula Planning District
  Commission
2017 Cunningham Drive
Hampton, VA 23666
  (804) 838-4238
  Director

Pennington County
Planning Department
County Courthouse
Thief River Falls, MN 56701

  (218) 681-2407
  Director

Pennington County
Planning and Zoning Commission
City Hall
22 Main Street
Rapid City, SD 57701
  (605) 394-2186
  Director: Frank McDaniel

Pennsylvania Association of Conservation District Directors
400 Fairview Avenue
Clarks Summit, PA 18411
  (717) 587-3346
  President: William Lange

Pennsylvania Citizens' Advisory
  Council
8th Floor
Executive House
P.O. Box 2357
Harrisburg, PA 17120
  (717) 787-4527
  Chairperson: Norman Childs

Pennsylvania
Department of Environmental Resources
Bureau of Environmental Planning
810 Executive House Apartments
Harrisburg, PA 17120
  (717) 783-1990
  Reviewer Liaison: David Blair

Pennsylvania
Department of Environmental Resources
Office of Environmental Protection
9th Floor Fulton Building
Harrisburg, PA 17120
  (717) 787-8104
  Reviewer Liaison: Jane Cico

Pennsylvania
Department of Environmental Resources
Office of Resources Management
230 Eban Press Building
Harrisburg, PA 17120
  (717) 787-2315
  Reviewer Liaison: Norm Kapko

Pennsylvania Environmental Council

225 South 15th Street
Philadelphia, PA   19102
   (215) 735-0966
   President: Curtin Winsor

Pennsylvania League of Cities
P.O. Box 5196
Harrisburg, PA   17110
   (717) 236-9469
   Executive Director: Richard Marden

Pennsylvania State Association of
   County Commissioners
301 Blackstone Building
112 Market Street
Harrisburg, PA   17101
   (717) 232-7554
   Executive Director: Jim Allen

Pepin County
Planning Department
County Courthouse
Durand, WI   54736
   (715) 672-8857
   Director

Perkins County
Planning Department
County Courthouse
Bison, SD   57620
   (605) 244-5626
   Director

Perquimans County
Planning Department
County Courthouse
Hertford, NC   27944
   (919) 426-5462
   Director

Perry County
Planning Department
County Courthouse
Cannelton, IN   47520
   (812) 547-3741
   Director

Person County
Planning Department
P.O. Box 1214
Roxboro, NC   27573
   (919) 599-9184

Director

Petroleum County
Planning Department
County Courthouse
Winnett, MT   59087
   (406) 429-5311
   Director

Phillips County
Planning Department
County Courthouse
Malta, MT   59538
   (406) 654-2423
   Director

Pico Rivera City
Planning Department
City Hall
Pico Rivera, CA   90660
   (213) 692-0401
   Director

Piedmont Planning District Commission
102½ High Street
P.O. Box P
Farmville, VA   23901
   (804) 392-6104
   Director

Piedmont Triad Council of Governments
2120 Pinecroft Road
Greensboro, NC   27407
   (919) 294-4950
   Regional Planning Director: Carl
   Loop

Pierce County
Planning Department
County Courthouse
Tacoma, WA   98402
   (206) 593-4000
   EIS Contact: Dan Cagle

Pierce County
Planning Department
County Courthouse
Ellsworth, WI   54011
   (715) 273-5272
   Director

Pike County

Planning Department
County Courthouse
Petersburg, IN 47567
  (812) 354-6025
  Director

Pine County
Planning Department
County Courthouse
Pine City, MN 55063
  (612) 629-3615
  Director

Pipestone County
Planning Department
County Courthouse
Pipestone, MN 56164
  (507) 825-4494
  Director

Pitt County
Planning Department
P.O. Box A
Greenville, NC 27834
  (919) 752-2934
  Director

Pittsfield City
Planning Department
City Hall
Pittsfield, MA 01201
  (413) 499-1100
  Director

Pittsylvania County
Planning Department
P.O. Box 426
Chatham, VA 24531
  (703) 432-2041
  County Administrator: William
    Sleeper

Placer County
Planning Department
175 Fulwelter
Auburn, CA 95603
  (916) 823-4381
  Director

Planning and Conservation League
717 K Street
Suite 209

Sacramento, CA 95814
  (916) 444-8726
  President: Thomas Ross

Plumas County
Planning Department
P.O. Box 207
Quincy, CA 95971
  (916) 283-1060
  Director

Plymouth County
Planning Department
P.O. Box 206
Plymouth, MA 02351
  (617) 746-4313
  Director

Polk County
Planning Department
County Courthouse
Crookston, MN 56716
  (218) 281-2332
  Director

Polk County
Planning Department
County Courthouse
Columbus, NC 28722
  (704) 894-3301
  Director

Polk County
Planning Department
County Courthouse
Balsam Lake, WI 54810
  (715) 485-3161
  County Planner: Brian O'Connell

Pomona City
Planning Department
City Hall
505 South Garey Avenue
Pomona, CA 91766
  (714) 620-2191
  Senior Planner: James Lightfoot

Pondera County
Planning Department
20 Fourth Avenue, S.W.
Conrad, MT 59425
  (406) 278-3021

Director

Pope County
Planning Department
County Courthouse
Glenwood, MN 56334
  (612) 634-3338
Director

Portage County
Planning Department
County Courthouse
Stevens Point, WI 54481
  (715) 346-2113
Director

Porter County
Planning Department
County Courthouse
Valparaiso, IN 46383
  (219) 462-3841
Director

Portsmouth City
Planning Department
P.O. Box 820
City Hall
Portsmouth, VA 23705
  (804) 393-8000
Director

Posey County
Planning Department
County Courthouse
Mount Vernon, IN 47620
  (812) 838-3492
Director

Potter County
Planning Department
County Courthouse
Gettysburg, SD 57442
  (605) 765-4461
Director

Powder River County
Planning Department
County Courthouse
Broadus, MT 59317
  (406) 436-2361
Director

Powell County
Planning Department
County Courthouse
Deer Lodge, MT 59722
  (406) 846-2772
Director

Powhatan County
Planning Department
P.O. Box 218
Powhatan, VA 23139
  (804) 598-3852
Director

Prairie County
Planning Department
County Courthouse
Terry, MT 59349
  (406) 637-5431
Director

Price County
Planning Department
County Courthouse
Phillips, WI 54555
  (715) 339-3325
Director

Prince Edward County
Planning Department
County Courthouse
Farmville, VA 23901
  (804) 392-4129
Director

Prince George County
Planning Department
County Courthouse
Upper Marlboro, MD 20870
  (301) 952-3397
Director

Prince George County
Planning Department
County Courthouse
Prince George, VA 23875
  (804) 732-8818
Director

Prince William County
Planning Department

9300-B Peabody Street
Manassas, VA 22110
  (703) 368-9171
  EIS Contact: Mukund Lokhande

Public Interest Research Group
1346 Connecticut Avenue, N.W.
Suite 415
Washington, DC 20036
  (202) 833-3934
  Director: Ralph Nader

Puerto Rico Association of Soil
  and Water Conservation Districts
P.O. Box 271
San German, PR 00753
  (809) 892-2860
  President: Eric Acosta

Puerto Rico
Environmental Quality Board
P.O. Box 11488
Santurce, PR 00910
  (809) 725-5140
  President: Pedro Gelabert

Puget Sound Council of Governments
216 First Avenue South
Seattle, WA 98104
  (206) 464-7090
  EIS Contact: Barbara Hastings

Pulaski County
Planning Department
County Courthouse
Winamac, IN 46996
  (219) 946-3653
  Director

Pulaski County
Planning Department
143 Third Street, N.W.
Pulaski, VA 24301
  (703) 980-8888
  Director

Putnam County
Planning Department
County Courthouse
Greencastle, IN 46135
  (317) 653-4019
  Director

Puttnam County
Planning Department
County Courthouse
Carmel, NY 10512
  (914) 225-3641
  Director

Queen Annes County
Planning Department
County Courthouse
Centreville, MD 21617
  (301) 758-0322
  Director

Quincy City
Planning Department
City Hall
Quincy, MA 02169
  (617) 773-1380
  Director

Rachel Carson Trust for the Living
  Environment
8940 Jones Mill Road
Washington, DC 20015
  (301) 652-1877
  President: Dr. Samuel Epstein

Racine City
Planning Department
City Hall
Racine, WI 53403
  (414) 636-9011
  Director

Racine County
Planning Department
County Courthouse
Racine, WI 53403
  (414) 636-3121
  Director

Raleigh City
Planning Department
110 South McDowell Street
P.O. Box 590
Raleigh, NC 27602
  (919) 755-6794
  EIS Contact: James Ratchford

Ramsey County
Department of Public Works

3377 North Rice Street
Saint Paul, MN 55112
  (612) 484-9104
  EIS Contact: Daniel Schacht

Rand Corporation
Office of the Vice President
1700 Main Street
Santa Monica, CA 90406
  (213) 393-0411
  Senior Vice President: Gustave
    Shubert

Randolph County
Planning Department
County Courthouse
Winchester, IN 47394
  (317) 584-7261
  Director

Randolph County
Planning Department
County Courthouse
Asheboro, NC 27203
  (919) 629-2131
  Director

Rappahannock Area Development Commission Planning District 16
1013 Princess Anne Street
Fredericksburg, VA 22401
  (703) 373-2890
  EIS Contact: Denise Butterfield

Rappahannock County
Planning Department
County Courthouse
Washington, VA 22747
  (703) 675-3621
  Director

Rappahannock-Rapidan Planning District Commission
125 West Locust Street
Culpeper, VA 22701
  (703) 825-6140
  Director

Ravalli County
Planning Department
County Courthouse
Hamilton, MT 59840

  (406) 363-1833
  Director

Red Lake County
Planning Department
County Courthouse
Red Lake Falls, MN 56750
  (218) 253-4281
  Director

Redondo Beach City
Planning Department
P.O. Box 270
Redondo Beach, CA 90277
  (213) 372-1171
  Director

Redwood City
Planning Department
P.O. Box 391
Redwood City, CA 94064
  (415) 369-6251
  Director

Redwood County
Planning Department
P.O. Box 70
Redwood Falls, MN 56283
  (507) 637-8325
  Director

Regional Planning Agency of South
  Central Connecticut
Regional Council of Elected Officials
  of South Central Connecticut
96 Grove Street
New Haven, CT 06510
  (203) 777-4795
  Director

Region L Council of Governments
P.O. Drawer 2748
Rocky Mount, NC 27801
  (919) 446-0411
  EIS Contact: Dwight Lamm

Renewable Natural Resources Foundation
5400 Grosvenor Lane
Washington, DC 20014
  (301) 897-8720
  Executive Director: Albert McClure

Rensselaer County
Planning Department
County Courthouse
Troy, NY 12180
  (518) 270-5220
  Director

Renville County
Planning Department
County Courthouse
Olivia, MN 56277
  (612) 523-2071
  Director

Resources for the Future
Institutions and Public Decisions
  Division
1755 Massachusetts Avenue, N.W.
Washington, DC 20036
  (202) 462-4400
  Tony Pryor

Rhode Island
Department of Environmental Management
83 Park Street
Providence, RI 02903
  (401) 277-2771
  Director: W. Edward Wood

Rhode Island League of Cities and
  Towns
39 Pike Street
Providence, RI 02903
  (401) 272-3434
  Executive Director: Ken Payne

Rhode Island State Association of
  Conservation Districts
Tourtellot Hill Road
Chepachet, RI 02814
  (401) 568-7373
  President: Domenic Marietti

Rice County
Planning Department
County Courthouse
Faribault, MN 55021
  (507) 334-2281
  Director

Richland County

Planning Department
County Courthouse
Sidney, MT 59270
  (406) 482-1708
  Director

Richland County
Planning Department
County Courthouse
Richland Center, WI 53581
  (608) 647-2747
  Director

Richmond City
Planning Department
City Hall
Richmond, CA 94804
  (415) 231-2080
  EIS Contact: Nancy Kaufman

Richmond City
Planning Department
City Hall
Richmond, VA 23219
  (804) 780-4000
  Director

Richmond County
Planning Department
County Courthouse
Rockingham, NC 28379
  (919) 997-2542
  Director: Michael Gurnee

Richmond County
Planning Department
County Courthouse
Warsaw, VA 22572
  (804) 333-8681
  Director

Richmond Regional Planning District
  Commission
6 North Sixth Street
Suite 500
Richmond, VA 23219
  (804) 644-8586
  EIS Contact: Timothy McGarry

Ripley County
Planning Department
County Courthouse

Versailles, IN 47042
  (812) 689-6115
  Director

River Hills Regional Planning
  Commission
c/o Indiana University
S.E. 4201 Grantline Road
New Albany, IN 47150
  (812) 945-2731
  Director

Riverside City
Planning Department
City Hall
Riverside, CA 92522
  (714) 787-7557
  Principal Planner: Steve Whyld

Riverside County
Planning Department
Environmental Quality Section
County Administrative Center
Ninth Floor
4080 Lemon Street
Riverside, CA 92501
  (714) 787-2331
  Director

Roanoke City
Planning Department
City Hall
Roanoke, VA 24011
  (703) 981-2333
  Director

Roanoke County
Planning Department
P.O. Box 168
Salem, VA 24153
  (703) 389-0811
  Director

Roberts County
Planning Department
County Courthouse
Sisseton, SD 57262
  (605) 698-3395
  Director

Robeson County
Planning Department

County Courthouse
Lumberton, NC 28358
  (919) 738-9341
  Director

Rochester City
Planning Department
City Hall
Rochester, MN 55901
  (507) 282-9495
  Director

Rochester City
Planning Department
City Hall
30 Church Street
Rochester, NY 14614
  (716) 428-7048
  Environmental Planner: Neil
    Freeland

Rochester-Olmsted Council of
  Governments
1421 Third Avenue, S.E.
Rochester, MN 55901
  (507) 285-8236
  Director

Rockbridge County
Planning Department
County Courthouse
Lexington, VA 24450
  (703) 463-4361
  Director

Rock County
Planning Department
County Courthouse
Luverne, MN 56156
  (507) 283-8212
  Director

Rock County
Planning Department
County Courthouse
51 South Main Street
Janesville, WI 53545
  (608) 755-2087
  EIS Contact: Phil Blazkowski

Rockingham County
Planning Department

P.O. Box 23
Wentworth, NC 27375
  (919) 349-2922
  Director

Rockingham County
Planning Department
County Courthouse
Harrisonburg, VA 22801
  (703) 434-5941
  Director

Rockland County
Planning Department
County Office Building
New City, NY 10956
  (914) 638-0500
  Director

Rocky Mountain Center on the
  Environment
1115 Grant Street
Denver, CO 80203
  President: J. Paul Heffron

Rome City
Planning Department
City Hall
Rome, NY 13440
  (315) 336-6000
  Director

Roosevelt County
Planning Department
County Courthouse
Wolf Point, MT 59201
  (406) 653-1322
  Director

Roseau County
Planning Department
County Courthouse
Roseau, MN 56751
  (218) 463-2541
  Director

Rosebud County
Planning Board
County Courthouse
Forsyth, MT 59327
  (406) 356-7551
  Coordinator: Eldon Rice

Rowan County
Planning Department
202 North Main Street
Salisbury, NC 28144
  (704) 636-0361
  Director

Rush County
Planning Department
County Courthouse
Rushville, IN 46173
  (317) 932-2086
  Director

Rusk County
Planning Department
County Courthouse
Ladysmith, WI 54848
  (715) 532-5556
  Director

Russell County
Planning Department
County Courthouse
Lebanon, VA 24266
  (703) 889-1931
  Director

Rutherford County
Planning Department
County Courthouse
Rutherfordton, NC 28139
  (704) 286-9136
  Director

Sacramento City
Planning Commission
City Hall
Sacramento, CA 95814
  (916) 449-5604
  EIS Contact: Clif Carstens

Sacramento County
Planning Department
827 Seventh Street
Room 101
Sacramento, CA 95814
  (916) 440-7914
  Environmental Coordinator: Alcides
    Freitas

Sacramento Regional Area Planning
  Commission

800 H Street
P.O. Box 808
Sacramento, CA 95804
  (916) 441-5930
  Director

Saint Cloud Area Council of Governments
46 North 28th Avenue
Saint Cloud, MN 56301
  (612) 252-7568
  EIS Contact: William Hansen

Saint Croix County
Planning Office
County Courthouse
Hudson, WI 54016
  (715) 386-5581
  County Planner: Richard Thompson

Saint Joseph County
Planning Department
County City Building
South Bend, IN 46601
  (219) 284-9011
  Director

Saint Lawrence County
Planning Department
County Courthouse
Canton, NY 13617
  (315) 379-2237
  Associate Planner: Jon Montan

Saint Louis County
Planning Department
County Courthouse
Duluth, MN 55802
  (218) 723-3521
  Director

Saint Marys County
Planning Department
County Courthouse
Leonardtown, MD 20650
  (301) 475-5621
  Director

Saint Paul City
Planning Department
25 West Fourth Street
Saint Paul, MN 55102

(612) 298-4012
EIS Contact: Rick Wiederhorn

Salinas City
Department of Community Development
200 Lincoln Avenue
Salinas, CA 93901
  (408) 758-7206
  Director: Roger Anderman

Sampson County
Planning Department
East Rowan Street
Box 303-C
Clinton, NC 28328
  (919) 592-6308
  EIS Contact: Neil Polansky

San Benito County
Planning Department
County Courthouse
Hollister, CA 95023
  (408) 637-3786
  Director

San Bernardino City
Planning Department
P.O. Box 1312
San Bernardino, CA 92401
  (714) 383-5002
  Director

San Bernardino County
Planning Department
1111 East Mill Street
Building 1
San Bernardino, CA 92415
  (714) 383-1417
  EIS Contact: Frederic Hinshaw

Sanborn County
Planning Department
County Courthouse
Woonsocket, SD 57385
  (605) 796-4513
  Director

San Buenaventura City
Planning Department
P.O. Box 99
San Buenaventura, CA 93001
  (805) 648-7881

EIS Contact: Susa Gates

Sanders County
Planning Department
County Courthouse
Thompson Falls, MT 59873
  (406) 827-3491
  Director

San Diego City
Planning Department
202 C Street
Fifth Floor
San Diego, CA 92101
  (714) 236-6363
  Associate Planner: Louis Tucker

San Diego County
Planning Department
County Administration Center
Room 207
1600 Pacific Highway
San Diego, CA 92101
  (714) 236-4597
  EIS Contact: Randall Hurlburt

San Francisco City
Planning Department
Office of Environmental Review
45 Hyde Street
Room 319
San Francisco, CA 94102
  (415) 558-3118
  EIS Contact: Gerald Owyang

San Francisco County
Planning Department
City Hall
San Francisco, CA 94102
  (415) 558-6161
  Director

San Joaquin County
Planning Department
1860 East Hazelton Avenue
Stockton, CA 95205
  (209) 944-2481
  EIS Contact: Gordon Moore

San Joaquin County Council of
  Governments
1850 East Hazelton

Stockton, CA 92505
  (209) 944-2585
  Director

San Jose City
Planning Department
City Hall Annex
Fourth Floor
San Jose, CA 95110
  (408) 277-4000
  Principal Planner: William Thomas

San Juan County
Planning Department
Box 947
Friday Harbor, WA 98250
  (206) 378-2354
  Director

San Leandro City
Planning Department
City Hall
San Leandro, CA 94577
  (415) 577-3000
  Director

San Luis Obispo County
Office of Environmental Coordinator
Courthouse Annex
San Luis Obispo, CA 93408
  (805) 549-5011
  Coordinator: Don Vossler

San Luis Obispo County and Cities Area
  Planning Coordinating Council
979 Osos Street
San Luis Obispo, CA 93401
  (805) 543-2550
  Director

San Mateo City
Department of Community Development
330 West 20th Avenue
San Mateo, CA 94403
  (415) 574-6770
  Environmental Planner: Jerome
    Podesta

San Mateo County
Planning Department
401 Marshall Street
Redwood City, CA 94063

(415) 364-5600
EIS Contact: Roman Gankin

Santa Ana City
Planning Department
20 Civic Center Plaza
P.O. Box 1988
Santa Ana, CA 92702
    (714) 834-4906
    Director: Chas. Zimmerman

Santa Barbara City
Planning Department
City Hall
Santa Barbara, CA 93102
    (805) 963-0611
    Director

Santa Barbara County
Planning Department
County Courthouse
Santa Barbara, CA 93104
    (805) 966-1611
    Director

Santa Barbara County-Cities Area
    Planning Council
1306 Santa Barbara Street
Santa Barbara, CA 93101
    (805) 966-1611
    Director

Santa Clara City
Planning Department
City Hall
Santa Clara, CA 93102
    (408) 984-3000
    Director

Santa Clara County
Planning Department
County Government Center
East Wing
San Jose, CA 95110
    (408) 299-2521
    EIS Contact: Richard Hall

Santa Cruz County
Planning Department
701 Ocean Street
Santa Cruz, CA 95062
    (408) 425-0111

Environmental Coordinator: Tom Burns

Santa Monica City
Planning Department
1685 Main Street
Santa Monica, CA 90401
    (213) 393-9975
    Director

Santa Rosa City
Department of Community Development
P.O. Box 1678
Santa Rosa, CA 95403
    (707) 528-5484
    EIS Contact: Marie Meredith

Saratoga County
Environmental Management Council
Municipal Center
Ballston Spa, NY 12020
    (518) 885-5381
    EIS Contact: George Hodgson, Jr.

Saulk County
Planning Department
County Courthouse
Boraboo, WI 53913
    (608) 356-5581
    Director

Save San Francisco Bay Association
P.O. Box 925
Berkeley, CA 94701
    (415) 849-3053
    President: William Siri

Sawyer County
Planning Department
County Courthouse
Hayward, WI 54843
    (715) 634-4866
    Director

Schenectady City
Planning Department
City Hall
Schenectady, NY 12305
    (518) 382-5111
    Director: Tom Macaulay

Schenectady County
Planning Department

620 State Street
Schenectady, NY 12307
  (518) 382-3286
  Senior Planner: G. David Foster

Schoharie County
Planning Department
County Building
Schoharie, NY 12157
  (518) 295-7147
  Director

Schuyler County
Planning Department
County Courthouse
Watkins Glen, NY 14891
  (607) 535-2132
  Director

Scientists' Institute for Public
  Information
355 Lexington Avenue
New York, NY 10017
  (212) 661-9110
  President: Alan McGowan

Scotland County
Planning Department
County Courthouse
Laurinburg, NC 28352
  (919) 276-3224
  Director

Scott County
Planning Department
County Courthouse
Scottsburg, IN 47170
  (812) 752-4769
  Director

Scott County
Planning Department
502 East First Avenue
Shakopee, MN 55379
  (612) 445-7750
  Director

Scott County
Planning Department
County Courthouse
Gale City, VA 24251
  (703) 386-6521

Director

Seacoast Anti-Pollution League
5 Market Street
Portsmouth, NH 03801
  (603) 431-5089
  President: Phil McDonough

Seattle City
Department of Community Development
SEPA Public Information Center
Yesler Building
400 Yesler Way
Seattle, WA 98104
  (206) 625-4537
  Supervisor: Trevor Evans

Seattle City
Executive Department
Office of Policy and Evaluation
400 Yesler Building
Fourth Floor
Seattle, WA 98104
  (206) 625-4575
  Land Use Planner: Rebecca Herzfeld

Seneca County
Planning Board
48 West Williams Street
Waterloo, NY 13165
  (315) 539-9285
  EIS Contact: Lawrence Mastrogiacomo

Shasta County
Planning Department
County Courthouse
Redding, CA 96001
  (916) 246-5631
  Environmental Coordinator: John
    Straham

Shawano County
Planning Department
County Courthouse
Shawano, WI 54166
  (715) 526-9150
  Director

Sheboygen County
Planning Department
County Courthouse
615 North 6th Street

Sheboygen, WI 53081
  (414) 459-3060
  County Planning Director: Mark
    Leider

Shelby County
Planning Department
County Courthouse
Shelbyville, IN 46176
  (317) 398-7448
  Director

Shenandoah County
Planning Department
P.O. Box 452
Woodstock, VA 22664
  (703) 459-2195
  Director

Sherburne County
Zoning Administration
Administration Building
328 Lowell Avenue
Elk River, MN 55330
  (612) 441-3160
  Zoning Administrator: Harvey
    Alfonds

Sheridan County
Planning Department
County Courthouse
Plentywood, MT 59254
  (406) 765-1660
  Director

Sibley County
Planning Department
County Courthouse
Gaylord, MN 55334
  (612) 237-2369
  Director

Sierra Club
530 Bush Street
San Francisco, CA 94108
  (415) 981-8634
  Executive Director: Michael
    McCloskey

Sierra Club
330 Pennsylvania Avenue, S.E.
Washington, DC 20003

  (202) 547-1144
  Director: Brock Evans

Sierra County
Planning Department
County Courthouse
Downieville, CA 95936
  (916) 289-3271
  Director

Sierra Planning Organization
11572 B Avenue
Auburn, CA 95603
  (916) 823-4703
  Director

Silver Bow County
Planning Department
County Courthouse
Butte, MT 59701
  (406) 792-3584
  Director

Simi Valley City
Planning Department
3200 Cochran Street
Simi Valley, CA 93065
  (805) 522-1333
  Director

Siskiyou Association of Governmental
  Entities
County Courthouse
Yreka, CA 96097
  (917) 842-3531
  Director

Siskiyou County
Planning Department
County Courthouse
Yreka, CA 96097
  (916) 842-3531
  Director

Skagit County
Planning Department
Room 218
County Administration Building
Mount Vernon, WA 98273
  (206) 336-9333
  Director: Robert Schofield

Skagit Regional Planning Council
120 West Kincaid Street
Mt. Vernon, WA 98273
   (206) 336-2188
   Director

Skamania County
Planning Department
County Courthouse
Stevenson, WA 98648
   (509) 427-5141
   Director

Skamania Regional Planning Council
20 Cascade Street
P.O. Box 152
Stevenson, WA 98648
   (509) 427-5141
   Director

Smithsonian Institution
1000 Jefferson Drive, S.W.
Washington, DC 20560
   (202) 628-4422
   Secretary: Dillion Ripley

Smyth County
Planning Department
P.O. Box 188
Marion, VA 24354
   (703) 783-3298
   Director

Snohomish County
Planning Department
County Courthouse
Everett, WA 98201
   (206) 259-9494
   Director

Society for Range Management
2760 West Fifth Avenue
Denver, CO 80204
   (303) 571-0174
   President: Daniel Merkel

Solano County
Planning Department
County Courthouse
Fairfield, CA 94533
   (707) 429-6412
   Director

Somerset County
Planning Department
County Courthouse
Princess Anne, MD 21853
   (301) 651-0320
   Director

Somerville City
Planning Department
City Hall
Somerville, MA 02143
   (617) 625-6600
   Director

Sonoma County
Planning Department
2555 Mendocino Avenue
Santa Rosa, CA 95401
   (707) 527-2241
   Director

Southampton County
Planning Department
County Courthouse
Courtland, VA 23837
   (804) 653-2465
   Director

South Bend City
Planning Department
County-City Building
South Bend, IN 46601
   (219) 284-9335
   Deputy Director: Steven Compton

South Carolina Association of Conser-
vation Districts
Box 414
Hartsville, SC 29550
   (803) 332-8151
   President: David Allen

South Carolina Association of Counties
1227 Main Street
808 SCN Center
Columbia, SC 29201
   (803) 252-7255
   Executive Director: Russell
      Shetterly

South Carolina
Department of Health and Environmental
   Control

Sims Building
2600 Bull Street
Columbia, SC 29201
    (803) 758-5654
    Commissioner: Albert Randall

South Carolina Municipal Association
P.O. Box 11558
Columbus, OH 29211
    (803) 799-9574
    Executive Director: McDonald Wray

South Dakota Association of County
    Commissioners
214 East Capitol
Pierre, SD 57501
    (605) 224-8654
    Executive Director: Neal Strand

South Dakota
Board of Environmental Protection
Room 408
Foss Building
Pierre, SD 57501
    (605) 773-3351
    Secretary: Allyn Lockner

South Dakota
Department of Agriculture
Conservation Division
Sigurd Anderson Building
Room 322
Pierre, SD 57501
    (605) 773-3258
    Director: Albert Griffiths

South Dakota
Department of Environmental Protection
Joe Foss Building
Pierre, SD 57501
    (605) 773-3351
    Acting Secretary: Richard Howard

South Dakota
Department of Natural Resource
    Development
Joe Foss Building
Pierre, SD 57501
    (605) 773-3151
    Secretary: Vern Butler

South Dakota Municipal League

214 East Capitol
Pierre, SD 57501
    (605) 224-8654
    Director: Robert Miller

South Dakota Planning and Development
    Commission
First District
401 First Avenue, N.E.
Watertown, SD 57201
    (605) 886-7224
    Director

South Dakota Planning and Development
    Commission
Third District
Yankton County Courthouse
P.O. Box 687
Yankton, SD 57078
    (605) 665-4408
    Director

South Dakota Planning and Development
    Commission
Fourth District
615 South Main
Aberdeen, SD 57401
    (605) 229-4740
    EIS Contact: Rick O'Connor

South Dakota Planning and Development
    Commission
Fifth District
365½ South Pierre Street
P.O. Box 640
Pierre, SD 57501
    (605) 224-1623
    Executive Director: Dennis Potter

South Dakota State Association of
    Conservation Districts
Oelrichs, SD 57763
    (605) 535-2045
    President: Vernon Seger

Southeastern Connecticut Regional
    Planning Agency
139 Boswell Avenue
Norwich, CT 06360
    (203) 889-2324
    Planner: Charles Storrow

South Eastern Council of Governments
208 East 13th Street
P.O. Box 1859
Sioux Falls, SD 57102
   (605) 336-1297
   Director

Southeastern Minnesota Regional
   Development Commission
301 Marquette Bank Building
Rochester, MN 55901
   (507) 285-2552
   Director

Southeastern Regional Planning and
   Economic Development District
7 Barnabas Road
Marion, MA 02738
   (617) 748-2100
   Executive Director: Alexander
      Zaleski

Southeastern Virginia Planning
   District Commission
16 Koeger Executive Center
Suite 100
Norfolk, VA 23502
   (804) 461-3200
   Executive Director: Arthur Collins

Southeastern Wisconsin Regional
   Planning Commission
916 North East Avenue
P.O. Box 769
Waukesha, WI 53187
   (414) 547-6721
   EIS Contact: Lyman Wible

Southern California Association
   of Governments
600 Commonwealth Avenue
Los Angeles, CA 90005
   (213) 385-1000
   Director

Southern Indiana Development
   Commission
P.O. Box 442
Loogootee, IN 47553
   (812) 295-3707
   Director

Southern Tier Central Regional
   Planning Development Board
53 Bridge Street
Corning, NY 14830
   (607) 962-5092
   Senior Environmental Planner:
      Jennifer Faig

Southern Tier East Regional Planning
   Development Board
Broome County Office Building
Government Plaza
P.O. Box 1766
Binghamton, NY 13902
   (607) 772-2856
   EIS Contact: Richard McCormick

Southern Tier West Regional Planning
   and Development Board
41 Main Street
Salamanca, NY 14779
   (716) 945-5303
   Director

South Gate City
Department of Community Development
8650 California Avenue
South Gate, CA 90280
   (213) 567-1331
   Senior Planner: Tim O'Rourke

Southside Planning District Commission
123 South Mecklenberg Avenue
P.O. Box 150
South Hill, VA 23970
   (804) 447-7101
   Director

Southwestern Indiana and Kentucky
   Regional Council of Governments
Civic Center Complex
Administration Building
Evansville, IN 47708
   (812) 426-5117
   Director

Southwestern North Carolina Planning
   and Economic Development Commission
P.O. Box 850
Bryson City, NC 28713
   (704) 488-2117
   Director

South Western Regional Planning
  Agency
137 Rowayton Avenue
Rowayton, CT 06853
  (203) 866-5543
  Director

Southwestern Wisconsin Regional
  Planning Commission
Pioneer Tower
Platteville, WI 53818
  (608) 342-1214
  Director

Southwest Regional Development
  Commission
P.O. Box 265
Slayton, MN 56172
  (507) 836-8549
  Executive Director: Jerry
    Chasteen

Spencer County
Planning Department
County Courthouse
Rockport, IN 47635
  (812) 649-4916
  Director

Spink County
Planning Department
County Courthouse
Redfield, SD 57469
  (605) 472-1825
  Director

Spokane City
Planning Commission
309 City Hall
Spokane, WA 99201
  (509) 456-3232
  Director: Terry Clegg

Spokane County
Planning Department
North 721 Jefferson Street
Spokane, WA 99260
  (509) 456-2205
  EIS Contact: Thomas Mosher

Spokane Regional Planning Conference
Spokane City Hall

Spokane, WA 99201
  (509) 456-4340
  Director

Sportsmen's Clubs of Texas
311 Vaughn Building
Austin, TX 78701
  (512) 472-2267
  President: Joe White

Spotsylvania County
Planning Department
P.O. Box 77
Spotsylvania, VA 22553
  (703) 582-6361
  Director

Springfield City
Planning Department
City Hall
Springfield, MA 01103
  (413) 736-2711
  Director

Stafford County
Planning Department
County Courthouse
Stafford, VA 22554
  (703) 659-4101
  Director

Stanford City
Planning Department
City Hall
Stanford, CT 06901
  (203) 348-5841
  Director

Stanislaus Area Association of
  Governments
814-14th Street
Modesto, CA 95354
  (209) 526-6200
  Director

Stanislaus County
Planning Department
P.O. Box 3404
Modesto, CA 95353
  (209) 526-6414
  Director

Stanley County
Planning Department
County Courthouse
Fort Pierre, SD 57532
　(605) 223-2642
　Director

Stanly County
Planning Department
County Courthouse
Albemarle, NC 28001
　(704) 983-2181
　Director

Starke County
Planning Commission
County Courthouse
Knox, IN 46534
　(219) 772-4688
　Executive Secretary: Art Hart

Statewide Program of Action to Con-
　serve Our Environment
Box 757
Concord, NH 03301
　(603) 679-8731
　Chairman: John Shortlidge

Stearns County
Planning Department
County Courthouse
Saint Cloud, MN 56301
　(612) 251-7833
　Director

Steele County
Planning Department
County Courthouse
Owatonna, MN 55060
　(507) 451-8040
　Director

Steuben County
Planning Department
County Courthouse
Angola, IN 46703
　(219) 665-3014
　Director

Steuben County
Environmental Management Board
21 East Morris

Bath, NY 14810
　(607) 776-2161
　Director: Nelson Parks

Stevens County
Planning Department
County Courthouse
Morris, MN 56267
　(612) 589-4660
　Director

Stevens County
Planning Department
County Courthouse
Colville, WA 99114
　(509) 684-4301
　Director

Stillwater County
Planning Department
Box 881
Columbus, MT 59019
　(406) 322-5328
　EIS Contact: Tom Kelly

Stockton City
Planning Department
City Hall
Stockton, CA 95202
　(209) 944-8459
　Director

Stokes County
Planning Department
County Courthouse
Danbury, NC 27016
　(919) 593-8161
　Director

Strategies for Environmental Control
631 West Main
Louisville, KY 40202
　(502) 587-3028
　President: Ralph Madison

Stratford City
Planning Department
City Hall
Stratford, CT 06497
　(203) 375-5621
　Director

Suffolk County
Planning Department
City Hall
Boston, MA 02201
  (617) 742-9250
  Director

Suffolk County
Council on Environmental Quality
Dennison Building
Veterans Memorial Highway
Hauppauge, NY 11787
  (516) 360-5204
  Principal Planner: James Bagg

Sullivan County
Planning Department
County Courthouse
Sullivan, IN 47882
  (812) 268-4657
  Director

Sullivan County
Planning Department
Government Center
Monticello, NY 12701
  (914) 794-3000
  Director

Sully County
Planning Department
County Courthouse
Onida, SD 57564
  (605) 258-2535
  Director

Sunnyvale City
Planning Department
City Hall
Sunnyvale, CA 94086
  (408) 739-0531
  Director

Surry County
Planning Department
P.O. Box 516
Dobson, NC 27017
  (919) 386-8676
    Associate County Manager: Dennis
      Thompson

Surry County

Planning Department
County Courthouse
Surry, VA 23883
  (804) 294-3266
  Director

Susquehanna River Basin Association
165 South Franklin Street
Wilkes-Barre, PA 18703
  (717) 824-5193
  President: Donald Smith

Sussex County
Planning Department
P.O. Box 1397
Sussex, VA 23884
  (804) 294-3266
  Director

Sutter County
Planning Department
463 Second Street
Yuba City, CA 95991
  (916) 673-5140
  Director

Swain County
Planning Department
Drawer A
Bryson City, NC 28713
  (704) 488-3121
  Director

Sweet Grass County
Planning Department
County Courthouse
Big Timber, MT 59011
  (406) 932-2713
  Director

Swift County
Planning Department
County Courthouse
Benson, MN 56215
  (612) 842-6271
  Director

Switzerland County
Planning Department
County Courthouse
Vevay, IN 47043
  (812) 427-3175

Director

Syracuse City
Department of Community Development
211 Montgomery Street
Hills Building
Syracuse, NY 13202
 (315) 473-2873
 EIS Contact: Terry Tipple

Tacoma City
Planning Department
County-City Building
Tacoma, WA 98402
 (206) 593-4411
 Director

Tahoe Regional Planning Agency
P.O. Box 8896
South Lake Tahoe, CA 95731
 (916) 541-0246
 Executive Director: Philip
  Overeynder

Talbot County
Planning Department
County Courthouse
Easton, MD 21601
 (301) 822-2401
 Director

Taylor County
Planning Department
County Courthouse
Medford, WI 54451
 (715) 748-3131
 Director

Tazewell County
Planning Department
County Courthouse
Tazewell, VA 24615
 (703) 988-5962
 Director

Tehama County
Planning Department
County Courthouse
Red Bluff, CA 96080
 (916) 527-3563
 Director

Tennessee Association of Conservation
 Districts
128 North Alpine
Ripley, TN 38063
 (901) 635-1407
 President: Talmadge Crihfield

Tennessee
Bureau of Environmental Health
 Services
349 Cordell Hull Building
Nashville, TN 37219
 (615) 741-3657
 Commissioner for the Environment:
  Dr. Eugene Fowinkle

Tennessee Citizens for Wilderness
 Planning
130 Tabor Road
Oak Ridge, TN 37830
 (615) 482-2153
 President: Lynn Dye

Tennessee Conservation League
1720 West End Avenue
Suite 600
Nashville, TN 37203
 (615) 329-4230
 President: Bill Blackburn

Tennessee County Services Association
226 Capital Boulevard Building
Nashville, TN 37219
 (615) 242-5591
 Executive Director: Ralph Harris

Tennessee
Department of Conservation
2611 West End Avenue
Nashville, TN 37203
 (615) 741-2301
 Commissioner: B.R. Allison

Tennessee Environmental Council
P.O. Box 1422
Nashville, TN 37202
 (615) 251-1110
 President: Duncan Callicott

Tennessee Municipal League
Room 317

226 Capitol Boulevard
Nashville, TN 37219
  (615) 255-6416
  Executive Director: Herbert Bingham

Terre Haute City
Planning Department
City Hall
Terre Haute, IN 47808
  (812) 232-3375
  Director

Teton County
Planning Department
County Courthouse
Choteau, MT 59422
  (406) 466-2693
  Director

Texas Association of Counties
P.O. Box 2131
Austin, TX 78768
  (512) 478-8753
  Executive Director: Sam Clonts

Texas Committee on Natural Resources
6805 Hillcrest Avenue
Room 214
Dallas, TX 75205
  (214) 368-5976
  Chairman: Edward Fritz

Texas County Judges' and Commis‐
  sioners' Association
Medina County Courthouse
Hondo, TX 78861
  (512) 426-2352
  President: Jerome Decker

Texas
Guadalupe-Blanco River Authority
P.O. Box 271
Seguin, TX 78155
  (512) 379-5822
  General Manager: John Specht

Texas Municipal League
1020 Southwest Tower
Austin, TX 78701
  (512) 478-6601
  Executive Director: Richard Brown

Thomas Jefferson Planning District
  Commission
701 East High Street
Charlottesville, VA 22901
  (804) 977-2870
  EIS Contact: James Skove

Thorne Ecological Institute
2336 Pearl Street
Boulder, CO 80302
  (303) 443-7325
  Executive Director: David Zimmerman

Threshold
International Center for Environmental
  Renewal
Suite 113
1785 Massachusetts Avenue, N.W.
Washington, DC 20036
  (202) 265-0020
  President: Peter Freeman

Thurston Regional Planning Council
Building 1
2000 Lakeridge Drive, S.W.
Olympia, WA 98502
  (206) 753-8131
  EIS Contact: Jim Kramer

Tioga County
Planning Department
County Courthouse
Owego, NY 13827
  (607) 687-3633
  Director

Tippecanoe County
Area Plan Commission
20 North Third
Lafayette, IN 47901
  (317) 423-9242
  Executive Director: James Hewley

Tipton County
Planning Department
County Courthouse
Tipton, IN 46072
  (317) 675-2795
  Director

Todd County
Planning Department

County Courthouse
Long Prairie, MN   56347
  (612) 732-6181
  Director

Tompkins County
Planning Department
128 East Buffalo Street
Ithaca, NY   14850
  (607) 274-5431
  Commissioner: Frank Liguri

Toole County
Planning Department
County Courthouse
Shelby, MT   59474
  (406) 434-2232
  Director

Torrance City
Planning Department
3031 Torrance Boulevard
Torrance, CA   90503
  (213) 328-5310
  Planning Associate: Michael Bihn

Transylvania County
Planning Department
County Courthouse
Brevard, NC   28712
  (704) 883-9046
  Director

Traverse County
Planning Department
County Courthouse
Wheaton, MN   56296
  (612) 563-4242
  Director

Treasure County
Planning Department
County Courthouse
Hysham, MT   59038
  (406) 342-5547
  Director

Trempealeau County
Planning Department
County Courthouse
Whitehall, WI   54773
  (715) 538-4717

Director

Triangle J Council of Governments
P.O. Box 12276
100 Park Drive
Research Triangle Park, NC   27709
  (919) 549-0551
  Director: Raymond Green

Trico Economic Development District
401 North Wynne
P.O. Box 214
Colville, WA   99114
  (509) 684-4571
  Director

Tri-County Area Planning Council
c/o Glenn County Planning Department
525 West Sycamore Street
Willows, CA   95988
  (916) 934-3388
  Director

Tri-County Council for Southern
  Maryland
P.O. Box 301
Waldorf, MD   20601
  (301) 645-2693
  Director

Tri-County Council for Western
  Maryland
Algonquin Motor Inn
Cumberland, MD   21502
  (301) 722-6885
  Director

Trinity County
Planning Department
P.O. Box 936
Weaverville, CA   96093
  (916) 623-5594
  Director: Jeff Shields

Tripp County
Planning Department
County Courthouse
Winner, SD   57580
  (605) 842-3727
  Director

Tri-State Regional Planning Commission

One World Trade Center
82nd Floor
New York, NY 10048
   (212) 938-3300
   EIS Contact: Norma Hessie

Troy City
Planning Department
City Hall
Troy, NY 12180
   (518) 270-4000
   Director

Trustees for Alaska
825 D Street
Suite 202
Anchorage, AK 99501
   (907) 276-4244
   Executive Director: Wilson Rice

Trust for Public Land
Office of the President
82 Second Street
San Francisco, CA 94105
   (415) 495-4014
   Joel Kuperberg

Trust Territory of the Pacific
   Islands
Department of Health Services
Trust Territory of the Pacific
   Islands
Saipan, CM 96950
   Chief of Environmental Health:
   Nachsa Siren

Tulare County
Planning Department
County Courthouse
Visalia, CA 93277
   (209) 733-6271
   Director

Tulare County Association of
   Governments
Courthouse
Visalia, CA 93277
   (209) 733-6303
   Director

Tuolumne County
Planning Department

43 North Green Street
Sonora, CA 95370
   (209) 532-8151
   Director: James Nuzum

Turner County
Planning Department
County Courthouse
Parker, SD 57053
   (605) 297-3115
   Director

Tyrrell County
Planning Department
P.O. Box 121
Columbia, NC 27925
   (919) 796-5611
   Director

Ulster County
Planning Department
County Courthouse
Kingston, NY 12401
   (914) 331-9300
   Director

Union County
Planning Department
County Courthouse
Liberty, IN 47353
   (317) 458-5464
   Director

Union County
Planning Department
P.O. Box 218
Monroe, NC 28110
   (704) 289-5511
   Director

Union County
Planning Department
Box 519
Elk Point, SD 57025
   (605) 356-2132
   Director

Union Town
Planning Department
311 East Main Street
Endwell, NY 13760
   (607) 754-2102

EIS Contact: Nancy Berkowitz

United New Conservationists
P.O. Box 362
Cambell, CA 95008
   (408) 296-0943
   President: Lilyan Brannon

U.S.
Appalachian Regional Commission
1666 Connecticut Avenue, N.W.
Washington, DC 20235
   (202) 673-7869
   Federal Cochairman: Robert Scott

U.S.
Architect of the Capitol
United States Capitol
Room S8 15
Washington, DC 20515
   (202) 225-1200
   Architect: George White

U.S.
Civil Aeronautics Board
Office of Economic Analysis
Environmental Programs Division
Washington, DC 20428
   (202) 655-4000
   Chief: Arnold Konheim

U.S.
Coast Guard
Environmental Impact Branch
G-WEP-7/62
400-7th Street, S.W.
Washington, DC 20590
   (202) 426-3300
   Chief: Don Dumlao

U.S.
Consumer Product Safety Commission
Bureau of Economic Analysis
5401 Westbard Avenue
Bethesda, MD 20207
   (301) 492-6647
   Deputy Associate Executive Director
      for Economics: Walter Hobby

U.S.
Council on Environmental Quality
722 Jackson Place, N.W.

Washington, DC 20006
   (202) 395-5700
   General Counsel: Nick Yost

U.S.
Council on Environmental Quality
722 Jackson Place, N.W.
Washington, DC 20006
   (202) 633-6027
   Public Information Officer: Carol
      Isber

U.S.
Delaware River Basin Commission
18th and C Streets, N.W.
Washington, DC 20240
   (202) 343-5761
   Commissioner: Sherman Tribbert

U.S.
Department of Agriculture
Science and Education Administration
Environmental Quality
Washington, DC 20250
   (301) 344-3278
   Dr. Jesse Lunin

U.S.
Department of Commerce
Economic Development Administration
Washington, DC 20230
   (202) 377-5113
   Special Assistant for the Environ-
      ment: Andrew Kauders

U.S.
Department of Commerce
Economic Development Administration
Washington, DC 20230
   (202) 377-5113
   Director of Public Affairs: Arch
      Parsons

U.S.
Department of Commerce
Economic Development Administration
Suite 505
Title Building
909-17th Street
Denver, CO 80202
   (303) 837-3981
   Regional Environmentalist: Gordon
      Butcher

U.S.
Department of Commerce
Economic Development Administration
Suite 700
1365 Peachtree Street, N.W.
Atlanta, GA 30309
  (404) 881-7667
    Regional Environmentalist: John
    Cole

U.S.
Department of Commerce
Economic Development Administration
175 West Jackson Boulevard
Suite A-1630
Chicago, IL 60604
  (312) 353-9531
    Acting Regional Environmentalist:
    Mike Budarz

U.S.
Department of Commerce
Economic Development Administration
Federal Reserve Bank Building
105 North 7th Street
Room 600
Philadelphia, PA 19106
  (215) 597-7806
    Acting Regional Environmentalist:
    John Marshall

U.S.
Department of Commerce
Economic Development Administration
American Bank Tower
Suite 600
221 West Sixth Street
Austin, TX 78701
  (512) 397-5849
    Regional Environmentalist

U.S.
Department of Commerce
Economic Development Administration
Lake Union Building
Suite 500
1700 West Lake Avenue, North
Seattle, WA 98109
  (206) 442-1675
    Regional Environmentalist: Larry
    Burr

U.S.
Department of Commerce
Environmental Affairs
Washington, DC 20230
  (202) 377-4335
    Deputy Assistant Secretary: Sidney
    Galler

U.S.
Department of Commerce
NOAA
Environmental Research Laboratories
3100 Marine Avenue
Boulder, CO 80302
  (303) 449-1000
    Director

U.S.
Department of Commerce
NOAA
National Marine Fisheries Service
Office of Resource Conservation and
  Management
Page Building 2
Room 418
3300 Whitehaven Street, N.W.
Washington, DC 20235
  (202) 634-7218
    Director

U.S.
Department of Commerce
Office of Coastal Zone Management
3300 Whitehaven Street, N.W.
Washington, DC 20235
  (202) 634-4232
    Assistant Administrator: Robert
    Knecht

U.S.
Department of Commerce
NOAA
Office of Coastal Zone Management
NEPA Compliance Unit
3300 Whitehaven Street, N.W.
Washington, DC 20235
  (202) 634-4253
    Chief: Dr. Robert Kifer

U.S.
Department of Commerce
NOAA

Office of Ecology and Conservation
Room 5813
Washington, DC 20230
  (202) 377-5181
  Director: Joyce Wood

U.S.
Department of Commerce
Regional Action Planning Commission
Coastal Plains Region
Suite 413
1725 K Street, N.W.
Washington, DC 20006
  (202) 634-3910
  Federal Cochairman: Claud Anderson

U.S.
Department of Commerce
Regional Action Planning Commission
Four Corners Region
Room 1898-C
Commerce Building
Washington, DC 20230
  (202) 377-5534
  Federal Cochairman: Ken Baskette

U.S.
Department of Commerce
Regional Action Planning Commission
New England Region
2606 Commerce Building
Washington, DC 20230
  (202) 377-4343
  Federal Cochairman: Joseph
    Grandmaison

U.S.
Department of Commerce
Regional Planning Commission
Old West Region
Suite 426
1730 K Street, N.W.
Washington, DC 20006
  (202) 634-3907
  Federal Cochairman: George
    McCarthy

U.S.
Department of Commerce
Regional Action Planning Commission
Ozark Region
Room 2099-B

Commerce Building
Washington, DC 20230
  (202) 377-2572
  Federal Cochairman: Pat Danner

U.S.
Department of Commerce
Regional Action Planning Commission
Pacific Northwest Region
Suite 122
Hall of States
444 North Capitol Street, N.W.
Washington, DC 20001
  (202) 254-7030
  Federal Cochairman: Patrick Vaughan

U.S.
Department of Commerce
Regional Action Planning Commission
Southwest Border Region
Suite 306
1111-20th Street, N.W.
Washington, DC 20006
  (202) 634-3917
  Federal Cochairman: Cristobal
    Aldrete

U.S.
Department of Commerce
Regulatory Economics and Policy Office
Constitution Avenue at 14th Street,
  N.W.
Washington, DC 20230
  (202) 377-2482
  Director: Robert Miki

U.S.
Department of Defense
Defense Logistics Agency
Attn: DLA-WS
Room 4D489
Cameron Station
Alexandria, VA 22314
  (202) 325-0188
  Director

U.S.
Department of Defense
Environmental Policy
The Pentagon
Washington, DC 20301
  (202) 545-6700

Director: Col. Donald Sadler

U.S.
Department of Energy
Assistant Secretary for Environment
Washington, DC 20545
   (202) 376-4000
   Deputy: James Liverman

U.S.
Department of HEW
Office of Environmental Affairs
Room 524 F2
Humphrey Building
200 Independence Avenue, S.W.
Washington, DC 20201
   (202) 245-7243
   Director: Charles Custard

U.S.
Department of HEW
Office of Human Development Services
Room 302E
200 Independence Avenue, S.W.
Washington, DC 20201
   (202) 245-7000
   Warren Master

U.S.
Department of HEW
Public Affairs
200 Independence Avenue, S.W.
Washington, DC 20201
   (202) 245-6296
   Assistant Secretary: Eileen
      Shanahan

U.S.
Department of HUD
Community Planning and Development
451 Seventh Street, S.W.
Washington, DC 20410
   (202) 755-5111
   Assistant Secretary: Robert
      Embry, Jr.

U.S.
Department of HUD
Office of Architecture and Engineer-
   ing Standards
451 Seventh Street, S.W.
Washington, DC 20410

(202) 755-5111
Director: James McCullough

U.S.
Department of HUD
Urban Development
451 Seventh Street, S.W.
Washington, DC 20410
   (202) 755-5111
   Associate General Counsel: Robert
      Kenison

U.S.
Department of Justice
Land and Natural Resources Division
Constitution Avenue and 10th Street,
   N.W.
Washington, DC 20530
   (202) 633-2000
   Assistant Attorney General: James
      Moorman

U.S.
Department of Labor
Office of Environmental Impact
   Assessment
Room N-3673
Washington, DC 20210
   (202) 523-7111
   Chief: Joanne Lindhard

U.S.
Department of Labor
OSHA
Office of Regulatory Analysis
Washington, DC 20210
   (202) 523-8165
   Director: Mary Weber

U.S.
Department of State
Environmental and Population Affairs
Washington, DC 20520
   (202) 632-7964
   Deputy Assistant Secretary: William
      Hayne

U.S.
Department of State
Office of Environmental Affairs
Washington, DC 20520
   (202) 632-7964

Director: Donald King

U.S.
Department of the Air Force
Deputy for Environment and Safety
Room 4C885
The Pentagon
Washington, DC 20330
  (202) 697-0800
  Deputy: Dr. Carlos Stern

U.S.
Department of the Air Force
Environmental Planning Directorate
Air Force Engineering and Services
  Center
Tyndall AFB, FL 32403
  (904) 283-6182
  Captain Dwight Clark

U.S.
Department of the Air Force
Environmental Planning Division
Washington, DC 20330
  (202) 697-7799
  Chief: Col. Hisao Yamada

U.S.
Department of the Air Force
Environmental Policy and Assessment
  Branch
Washington, DC 20330
  (202) 695-2889
  Chief: Col. Francis Smith

U.S.
Department of the Army
Alaska District
P.O. Box 7002
Anchorage, AK 99510
  (907) 276-4915
  EIS Contact: William Lloyd

U.S.
Department of the Army
Army Corps of Engineers
Albuquerque District
P.O. Box 1580
Albuquerque, NM 87103
  (505) 766-2657
  EIS Contact: David Clawson

U.S.
Department of the Army
Army Corps of Engineers
Assistant Chief of Engineers
Environmental Office
HQDA(DAEN-ZCE)
Washington, DC 20310
  (202) 694-1163
  Col. Kenneth Halleran

U.S.
Department of the Army
Army Corps of Engineers
Baltimore District
P.O. Box 1715
Baltimore, MD 21203
  (301) 962-2558
  EIS Contact: Larry Lower

U.S.
Department of the Army
Army Corps of Engineers
Buffalo District
1776 Niagara Street
Buffalo, NY 14207
  (716) 876-5454
  EIS Contact: James Bennett

U.S.
Department of the Army
Army Corps of Engineers
Charleston District
P.O. Box 919
Charleston, SC 29402
  (803) 724-4258
  EIS Contact: John Carother

U.S.
Department of the Army
Army Corps of Engineers
Chicago District
219 South Dearborn Street
Chicago, IL 60604
  (312) 353-6509
  EIS Contact: Edward Hanses

U.S.
Department of the Army
Army Corps of Engineers
Detroit District
P.O. Box 1027
Detroit, MI 48231

(313) 226-6752
EIS Contact: Mary Ann Cooper

U.S.
Department of the Army
Army Corps of Engineers
Fort Worth District
P.O. Box 17300
Ft. Worth, TX 76102
  (817) 334-2095
  EIS Contact: L.E. Horsman

U.S.
Department of the Army
Army Corps of Engineers
Galveston District
P.O. Box 1229
Galveston, TX 77553
  (713) 763-1211
  EIS Contact: C.R. Harbaugh

U.S.
Department of the Army
Army Corps of Engineers
Huntington District
P.O. Box 2127
Huntington, WV 25721
  (304) 529-5636
  EIS Contact: William Sinozich

U.S.
Department of the Army
Army Corps of Engineers
Jacksonville District
P.O. Box 4970
Jacksonville, FL 32201
  (904) 791-3615
  EIS Contact: Moray Harrell

U.S.
Department of the Army
Army Corps of Engineers
Kansas District
700 Federal Building
Kansas City, MO 64106
  (816) 374-3672
  EIS Contact: James Taylor

U.S.
Department of the Army
Army Corps of Engineers
Little Rock District

P.O. Box 867
Little Rock, AR 72203
  (501) 378-5834
  EIS Contact: R.S. Durham

U.S.
Department of the Army
Army Corps of Engineers
Los Angeles District
P.O. Box 2711
Los Angeles, CA 90053
  (213) 688-5421
  EIS Contact: Ken Kules

U.S.
Department of the Army
Army Corps of Engineers
Louisville District
P.O. Box 59
Louisville, KY 40201
  (502) 684-5696
  EIS Contact: Frank Christ

U.S.
Department of the Army
Army Corps of Engineers
Lower Mississippi Valley Division
P.O. Box 80
Vicksburg, MS 39180
  (601) 634-5849
  EIS Contact: Hugh Holland, III

U.S.
Department of the Army
Army Corps of Engineers
Memphis District
668 Clifford Davis Federal Building
Memphis, TN 38103
  (901) 521-3857
  EIS Contact: Andrew Grosso

U.S.
Department of the Army
Army Corps of Engineers
Missouri River Division
P.O. Box 103
Downtown Station
Omaha, NE 68101
  (402) 221-7279
  EIS Contact: David Billman

U.S.

Department of the Army
Army Corps of Engineers
Mobile District
P.O. Box 2288
Mobile, AL 36628
  (205) 690-2727
  EIS Contact: James Hildreth

U.S.
Department of the Army
Army Corps of Engineers
Nashville District
P.O. Box 1070
Nashville, TN 37202
  (615) 251-5872
  EIS Contact: Gene Ottinger

U.S.
Department of the Army
Army Corps of Engineers
NEPA Counsel
HQDA (DAEN-CCJ)
Washington, DC 20314
  (202) 272-0034
  EIS Contact: Lance Wood

U.S.
Department of the Army
Army Corps of Engineers
New England Division
424 Trapelo Road
Waltham, MA 02154
  (617) 894-2400
  EIS Contact: David Dupee

U.S.
Department of the Army
Army Corps of Engineers
New Orleans District
P.O. Box 60267
New Orleans, LA 70160
  (504) 838-2518
  EIS Contact: Suzanne Hawes

U.S.
Department of the Army
Army Corps of Engineers
New York District
26 Federal Plaza
New York, NY 10007
  (212) 264-4662
  EIS Contact: Simeon Hook

U.S.
Department of the Army
Army Corps of Engineers
Norfolk District
803 Front Street
Norfolk, VA 23510
  (804) 441-3766
  EIS Contact: Karl Kuhlmann

U.S.
Department of the Army
Army Corps of Engineers
North Atlantic Division
90 Church Street
New York, NY 10007
  (212) 264-7814
  EIS Contact: Charles Stone

U.S.
Department of the Army
Army Corps of Engineers
North Central Division
536 South Clark Street
Chicago, IL 60605
  (312) 358-5469
  EIS Contact: Gene Flemming

U.S.
Department of the Army
Army Corps of Engineers
North Pacific Division
P.O. Box 2870
Portland, OR 97208
  (503) 221-3832
  EIS Contact: Art Gerlach

U.S.
Department of the Army
Army Corps of Engineers
Office of the Chief of Engineers
Directorate of Military Programs
HQDA (DAEN-MPZ-H)
Washington, DC 20314
  (202) 272-0391
  Leslie Savage, COL

U.S.
Department of the Army
Army Corps of Engineers
Office of the Chief of Engineers
HQDA (DAEN-CWR-P)
Washington, DC 20314

(202) 272-0121
EIS Contact: Richard Makinen

U.S.
Department of the Army
Army Corps of Engineers
Office of the Chief of Engineers
Environmental Programs
Forrestal Building
Washington, DC 20314
  (202) 693-7093
    Assistant Director: Lt. Col. George
    Boone

U.S.
Department of the Army
Army Corps of Engineers
Ohio River Division
P.O. Box 1159
Cincinnati, OH 45201
  (513) 684-3077
    EIS Contact: Jeremiah Parsons

U.S.
Department of the Army
Army Corps of Engineers
Omaha District
6014 USPO and Courthouse
Omaha, NE 68102
  (402) 221-4605
    EIS Contact: Richard Gorton

U.S.
Department of the Army
Army Corps of Engineers
Pacific Ocean Division
Building 230
Ft. Shafter, HI 96858
  (808) 438-2263
    EIS Contact: James Maragos

U.S.
Department of the Army
Army Corps of Engineers
Philadelphia District
U.S. Custom House
Second and Chesnut Street
Philadelphia, PA 19106
  (215) 597-4833
    EIS Contact: John Burnes

U.S.

Department of the Army
Army Corps of Engineers
Pittsburgh District
Federal Building
1000 Liberty Avenue
Pittsburgh, PA 15222
  (412) 644-6844
    EIS Contact: James Purdy

U.S.
Department of the Army
Army Corps of Engineers
Portland District
P.O. Box 2946
Portland, OR 97208
  (503) 221-6433
    EIS Contact: Lauren Aimonetto

U.S.
Department of the Army
Army Corps of Engineers
Regulatory Permits
HQDA (DAEN-CWO-N)
Washington, DC 20314
  (202) 272-0199
    EIS Contact: David Shepard

U.S.
Department of the Army
Army Corps of Engineers
Rock Island District
Clock Tower Building
Rock Island, IL 61201
  (309) 788-6361
    EIS Contact: Michael Cockerill

U.S.
Department of the Army
Army Corps of Engineers
Sacramento District
650 Capital Mall
Sacramento, CA 95814
  (916) 440-3120
    EIS Contact: Fred Kindel

U.S.
Department of the Army
Army Corps of Engineers
San Francisco District
211 Main Street
San Francisco, CA 94105
  (415) 556-8239

EIS Contact: Robert Schueneman

U.S.
Department of the Army
Army Corps of Engineers
Savannah District
P.O. Box 889
Savannah, GA 31402
  (912) 233-8822
  EIS Contact: Herbert Derigo

U.S.
Department of the Army
Army Corps of Engineers
Seattle District
P.O. Box C-3755
Seattle, WA 98124
  (206) 764-3624
  EIS Contact: Steven Dice

U.S.
Department of the Army
Army Corps of Engineers
South Atlantic Division
510 Title Building
30 Pryor Street, N.W.
Atlanta, GA 30303
  (404) 221-4331
  EIS Contact: John Rushing

U.S.
Department of the Army
Army Corps of Engineers
South Pacific Division
630 Sansome Street
Room 1216
San Francisco, CA 94111
  (415) 556-8775
  EIS Contact: James Sears

U.S.
Department of the Army
Army Corps of Engineers
Southwestern Division
Main Tower Building
1200 Main Street
Dallas, TX 75202
  (214) 767-2302
  EIS Contact: Water Gallaher

U.S.
Department of the Army

Army Corps of Engineers
St. Louis District
210 North 12th Street
St. Louis, MO 63101
  (314) 263-5711
  EIS Contact: Owen Dutt

U.S.
Department of the Army
Army Corps of Engineers
St. Paul District
1135 USPO and Custom House
St. Paul, MN 55101
  (612) 725-7632
  EIS Contact: Joseph Yanta

U.S.
Department of the Army
Army Corps of Engineers
Tulsa District
P.O. Box 61
Tulsa, OK 74102
  (918) 581-7858
  EIS Contact: Buell Atkins

U.S.
Department of the Army
Army Corps of Engineers
Vicksburg District
P.O. Box 60
Vicksburg, MS 39180
  (601) 634-5429
  EIS Contact: St. Clair Thompson

U.S.
Department of the Army
Army Corps of Engineers
Walla Walla District
Building 602
City-County Airport
Walla Walla, WA 99362
  (509) 525-5000
  EIS Contact: Ray Oligher

U.S.
Department of the Army
Army Corps of Engineers
Wilmington District
P.O. Box 1890
Wilmington, NC 28402
  (919) 343-4745
  EIS Contact: Richard Jackson

U.S.
Department of the Army
Army General Counsel
SAGC
Washington, DC 20310
   (202) 695-3306
   Cynthia Castle

U.S.
Department of the Army
Chief of Army Reserves
HQDA (DAAR-CM)
Washington, DC 20310
   (202) 697-2753
   Wil Gwilliam, LTC

U.S.
Department of the Army
Chief of Staff, Army
HQDA (DACS-DMA)
Washington, DC 20310
   (202) 694-4160
   Lewis Johnson, LTC

U.S.
Department of the Army
Deputy Assistant Secretary of the
   Army
Environment, Safety and Occupational
   Health
Secretary of the Army
Installations, Logistics and
   Financial Management
Washington, DC 20310
   (202) 695-7824
   Lewis Walker

U.S.
Department of the Army
Environmental Affairs
The Pentagon
Washington, DC 20310
   (202) 695-1370
   Deputy: Bruce Hilderbrand

U.S.
Department of the Army
Intelligence
HQDA (DAMI-TSP)
Washington, DC 20310
   (202) 695-4221
   Assistant Chief: James Beck

U.S.
Department of the Army
Judge Advocate General
HQDA (JALS-RL)
Washington, DC 20310
   (703) 756-2015
   Ernest Pearson, CPT

U.S.
Department of the Army
Logistics
HQDA (DALO-TSE)
Washington, DC 20310
   (202) 697-8503
   Deputy Chief: Roy Paul, LTC

U.S.
Department of the Army
National Guard Bureau
HQDA (NGB-ARI)
Washington, DC 20310
   (202) 697-3688
   Edward Huempfner

U.S.
Department of the Army
Office of the Assistant Chief of
   Engineers
Headquarters DAEN-ZCE
Environmental Office
Room 1E676
The Pentagon
Washington, DC 20310
   (202) 694-4269
   Chief: Col. Charles Sell

U.S.
Department of the Army
Personnel
HQDA (DAPE-MCP)
Washington, DC 20310
   (202) 695-5022
   Deputy Chief: Al Shauf, LTC

U.S.
Department of the Army
Plans and Operations
HQDA (DAMO-TRS)
Washington, DC 20310
   (202) 694-5700
   Deputy Chief: Matthew Jones, LTC

U.S.
Department of the Army
Public Affairs
HQDA (SAPA-PP)
Washington, DC 20310
  (202) 695-4462
    Chief: Warren Lacy, MAJ

U.S.
Department of the Army
Research, Development and Acquisition
HQDA (DAMA-AR)
Washington, DC 20310
  (202) 695-1449
    Deputy Chief: Charles Church

U.S.
Department of the Interior
Bureau of Land Management
18th and C Streets, N.W.
Washington, DC 20240
  (202) 343-5717
    Director: Frank Gregg

U.S.
Department of the Interior
Bureau of Land Management
Alaska Field Office
701 C Street
Box 13
Anchorage, AK 99513
  (907) 271-5076
    Director

U.S.
Department of the Interior
Bureau of Land Management
Arizona Field Office
2400 Valley Bank Center
Phoenix, AZ 85073
  (602) 261-3873
    Director

U.S.
Department of the Interior
Bureau of Land Management
California Field Office
Federal Building
Sacramento, CA 95825
  (916) 484-4676
    Director

U.S.
Department of the Interior
Bureau of Land Management
Colorado Field Office
Colorado State Bank Building
Denver, CO 80202
  (303) 837-4325
    Director

U.S.
Department of the Interior
Bureau of Land Management
Eastern States Field Office
7981 Eastern Avenue
Silver Spring, MD 20910
  (301) 427-7500
    Director

U.S.
Department of the Interior
Bureau of Land Management
Idaho Field Office
Federal Building
Boise, ID 83724
  (208) 384-1401
    Director

U.S.
Department of the Interior
Bureau of Land Management
Montana Field Office
Granite Tower Building
222 North 32nd Street
Billings, MT 59101
  (406) 657-6461
    Director

U.S.
Department of the Interior
Bureau of Land Management
Nevada Field Office
Federal Building
Reno, NV 89509
  (702) 784-5451
    Director

U.S.
Department of the Interior
Bureau of Land Management
New Mexico Field Office
Federal Building
Sante Fe, NM 87501

(505) 988-6217
Director

U.S.
Department of the Interior
Bureau of Land Management
Office of Public Affairs
Washington, DC 20240
    (202) 343-1100
    Chief: Daniel Alfieri

U.S.
Department of the Interior
Bureau of Land Management
Oregon Field Office
729 N.E. Oregon Street
Portland, OR 97208
    (503) 231-6251
    Director

U.S.
Department of the Interior
Bureau of Land Management
Utah Field Office
University Club Building
136 East South Temple Street
Salt Lake City, UT 84111
    (801) 524-5311
    Director

U.S.
Department of the Interior
Bureau of Land Management
Wyoming Field Office
Federal Building
Cheyenne, WY 82001
    (307) 778-2326
    Director

U.S.
Department of the Interior
Bureau of Mines
Washington, DC 20241
    (202) 634-1004
    Director: Roger Markle

U.S.
Department of the Interior
Bureau of Mines
2401 E Street, N.W.
Washington, DC 20241
    (202) 634-1325

EIS Preparation: Dr. Edwin Maust

U.S.
Department of the Interior
Bureau of Mines
Mineral Information Office
2401 E Street, N.W.
Washington, DC 20241
    (202) 634-1004
    Director

U.S.
Department of the Interior
Bureau of Reclamation
Engineering and Research Center
Building 67
Denver Federal Center
Denver, CO 80225
    (303) 234-2050
    Environmental Specialist: John
      Peters

U.S.
Department of the Interior
Bureau of Reclamation
Lower Colorado Region
Nevada Highway and Park Street
Box 427
Boulder City, NV 89005
    (702) 293-7411
    Director

U.S.
Department of the Interior
Bureau of Reclamation
Lower Missouri Region
Building 20
Box 25247
Denver Federal Center
Denver, CO 80225
    (303) 234-4441
    Director

U.S.
Department of the Interior
Bureau of Reclamation
Mid-Pacific Region
Federal Office Building
2800 Cottage Way
Sacramento, CA 95825
    (916) 484-4571
    Director

U.S.
Department of the Interior
Bureau of Reclamation
Office of Public Affairs
Washington, DC 20240
    (202) 343-4662
    Director

U.S.
Department of the Interior
Bureau of Reclamation
Pacific Northwest Region
550 West Fort Street
Box 043
Boise, ID 83724
    (208) 342-2101
    Director

U.S.
Department of the Interior
Bureau of Reclamation
Southwest Region
317 East Third
Box H-4377
Amarillio, TX 79101
    (806) 376-2401
    Director

U.S.
Department of the Interior
Bureau of Reclamation
Upper Colorado Region
125 South State
Box 11568
Salt Lake City, UT 84147
    (801) 524-5592
    Director

U.S.
Department of the Interior
Bureau of Reclamation
Upper Missouri Region
316 North 26th Street
Box 2553
Billings, MT 59103
    (406) 245-6214
    Director

U.S.
Department of the Interior
Conservation and Wildlife
Interior Building

C Street Between 18th and 19th
    Streets, N.W.
Washington, DC 20240
    (202) 343-1100
    Associate Solicitor: James Webb

U.S.
Department of the Interior
Environmental Affairs Office
Building 67
Denver Federal Center
Denver, CO 80225
    (303) 234-2050
    Chief: John Peters

U.S.
Department of the Interior
Environmental Project Review Office
Room 4256
Interior Building
Washington, DC 20240
    (202) 343-3891
    Director: Bruce Blanchard

U.S.
Department of the Interior
Environmental Project Review Office
Alburquerque Region
5301 Central Avenue, N.E.
Alburquerque, NM 87103
    (505) 474-3565
    Director

U.S.
Department of the Interior
Environmental Project Review Office
Anchorage Region
1675 C Street
Anchorage, AK 99510
    (907) 265-5278
    Director

U.S.
Department of the Interior
Environmental Project Review Office
Atlanta Region
148 International Boulevard, N.E.
Atlanta, GA 30303
    (404) 242-4524
    Director

U.S.

Department of the Interior
Environmental Project Review Office
Boston Region
15 State Street
Boston, MA 02109
(617) 223-5517
Director

U.S.
Department of the Interior
Environmental Project Review Office
Chicago Region
2510 Dempster Street
Chicago, IL 60016
(312) 298-2624
Director

U.S.
Department of the Interior
Environmental Project Review Office
Denver Region
Denver Federal Center
Denver, CO 80225
(303) 234-2071
Director

U.S.
Department of the Interior
Environmental Project Review Office
Portland Region
500 N.E. Mutnomah Street
Portland, OR 93272
(503) 429-6157
Director

U.S.
Department of the Interior
Environmental Project Review Office
San Francisco Region
450 Golden Gate Avenue
San Francisco, CA 94102
(415) 556-8200
Director

U.S.
Department of the Interior
Geological Survey
Central Region
Box 25046
Denver Federal Center
Denver, CO 80225
(303) 234-4630

Director

U.S.
Department of the Interior
Geological Survey
Eastern Region
109 National Center
Reston, VA 22092
(703) 860-7414
Director

U.S.
Department of the Interior
Geological Survey
National Cartographic Information
Center
CTR/507 National Center
Reston, VA 22092
(703) 860-6045
Director

U.S.
Department of the Interior
Geological Survey
National Center
12201 Sunrise Valley Drive
Reston, VA 22092
(703) 860-7444
Director

U.S.
Department of the Interior
Geological Survey
Western Region
345 Middlefield Road
Menlo Park, CA 94025
(415) 323-2711
Director

U.S.
Department of the Interior
Lower Colorado Region
P.O. Box 427
Boulder City, NV 89005
(702) 293-7560
Regional Environmental Affairs
Officer: Ken Trompeter

U.S.
Department of the Interior
Lower Missouri Region
Building 20

Denver Federal Center
Denver, CO 80225
  (303) 234-3779
    Regional Environmental Affairs
      Officer: Richard Eggen

U.S.
Department of the Interior
Mid-Pacific Region
2800 Cottage Way
Sacramento, CA 95825
  (916) 484-4792
    Regional Environmental Affairs
      Officer: Charles Long

U.S.
Department of the Interior
Office of Environmental Affairs
18th and C Streets, N.W.
Washington, DC 20240
  (202) 343-2840
    Director: Al R. Jonez

U.S.
Department of the Interior
Pacific Northwest Region
P.O. Box 043
U.S. Court House
550 West Fort Street
Boise, ID 83724
  (208) 384-1207
    Regional Environmental Affairs
      Officer: John Woodworth

U.S.
Department of the Interior
Southwest Region
Herring Plaza
Box H-4377
Amarillo, TX 79101
  (806) 376-2404
    Regional Environmental Affairs
      Officer: Alfred Hill

U.S.
Department of the Interior
Upper Colorado Region
P.O. Box 11568
Salt Lake City, UT 84111
  (801) 524-5580
    Regional Environmental Affairs
      Officer: Harold Sersland

U.S.
Department of the Interior
Upper Missouri Region
P.O. Box 2553
Billings, MT 59103
  (406) 657-6558
    Regional Environmental Affairs
      Officer: Eley Denson

U.S.
Department of the Navy
Naval Facilities Engineering Command
Natural Resources Program
200 Stoval Street
Alexandria, VA 22332
  (202) 325-0427
    Director: Harold Richard

U.S.
Department of the Navy
Office of the Chief of Naval Oper-
  ations
Environmental Impact Statement/RDT&E
  Branch
Washington, DC 20350
  (202) 697-3689
    Head: Ed Johnson

U.S.
Department of the Treasury
Washington, DC 20220
  (202) 566-2000
    Environmental Quality Officer:
      W.J. McDonald

U.S.
Department of the Treasury
Office of Administrative Programs
Environmental Programs
Washington, DC 20220
  (202) 376-0289
    NEPA Liaison: Anthony DiSilvestre

U.S.
Department of the Treasury
Office of the General Counsel
Administration and General Law
  Section
Washington, DC 20220
  (202) 566-2327
    NEPA Contact: Miklos Lonkay

U.S.
Environmental Protection Agency
Office of Enforcement
Waterside Mall West Tower
Room 1111
Washington, DC 20460
  (202) 755-2500
    Assistant Administrator: Marvin
      Durning

U.S.
Environmental Protection Agency
Office of Enforcement
General Enforcement
Waterside Mall West Tower
Room 1111
Washington, DC 20460
  (202) 755-2977
    Deputy Assistant Administrator:
      Richard Wilson

U.S.
Environmental Protection Agency
Office of Enforcement
General Enforcement
Pesticides and Toxic Substances
Waterside Mall
401 M Street, S.W.
Room 3624
Washington, DC 20460
  (202) 755-0970
    Director: Augustine Conroy, II

U.S.
Environmental Protection Agency
Office of Enforcement
General Enforcement
Stationary Source
Waterside Mall
401 M Street, S.W.
Room 3202
Washington, DC 20460
  (202) 755-2550
    Director: Edward Reich

U.S.
Environmental Protection Agency
Office of Enforcement
Mobile Source
Waterside Mall
401 M Street, S.W.
Room 3220

Washington, DC 20460
  (202) 755-2870
    Director: Charles Freed

U.S.
Environmental Protection Agency
Office of Enforcement
Source and Noise Enforcement
Waterside Mall West Tower
Room 1111
Washington, DC 20460
  (202) 755-2530
    Deputy Assistant Administrator:
      Ben Jackson

U.S.
Environmental Protection Agency
Office of Enforcement
National Enforcement Investigation
  Center
Building 53
Box 25227
Denver Federal Center
Denver, CO 80225
  (303) 234-4650
    Director: Thomas Gallagher

U.S.
Environmental Protection Agency
Office of Enforcement
Water Enforcement
Waterside Mall West Tower
Room 1119
Washington, DC 20460
  (202) 755-0440
    Deputy Assistant Administrator:
      Jeffrey Miller

U.S.
Environmental Protection Agency
Office of Environmental Review
Waterside Mall
401 M Street, S.W.
Washington, DC 20460
  (202) 755-0777
    Director: William Hedeman, Jr.

U.S.
Environmental Protection Agency
Office of Environmental Review
Community Development Liaison Staff
Waterside Mall

401 M Street, S.W.
Washington, DC 20460
  (202) 755-0780
    Assistant Director: Joseph McCabe

U.S.
Environmental Protection Agency
Office of Environmental Review
Policy and Procedures Staff
Waterside Mall
401 M Street, S.W.
Washington, DC 20460
  (202) 755-0790
    Assistant Director: Peter Cook

U.S.
Environmental Protection Agency
Office of Environmental Review
Resource Development Staff
Waterside Mall
401 M Street, S.W.
Washington, DC 20460
  (202) 755-0770
    Assistant Director: William
    Dickerson

U.S.
Environmental Protection Agency
Office of General Counsel
Waterside Mall West Tower
Room 1037D
Washington, DC 20460
  (202) 755-2511
    General Counsel: Joan Bernstein

U.S.
Environmental Protection Agency
Office of General Counsel
Air Quality, Noise and Radiation
  Division
Waterside Mall West Tower
Room 545
Washington, DC 20460
  (202) 755-0744
    Associate General Counsel: Michael
    James

U.S.
Environmental Protection Agency
Office of General Counsel
Pesticides Division
Waterside Mall West Tower

Room 537
Washington, DC 20460
  (202) 755-2680
    Associate General Counsel: David
    Menotti

U.S.
Environmental Protection Agency
Office of General Counsel
Toxic Substances Division
Waterside Mall West Tower
Room 537
Washington, DC 20460
  (202) 755-0794
    Associate General Counsel: Richard
    Denney

U.S.
Environmental Protection Agency
Office of General Counsel
Water and Solid Waste Division
Waterside Mall West Tower
Room 511
Washington, DC 20460
  (202) 755-0753
    Associate General Counsel: James
    Rogers

U.S.
Environmental Protection Agency
Office of Legislation
Waterside Mall West Tower
Room 835
Washington, DC 20460
  (202) 755-2930
    Director: Charles Warren

U.S.
Environmental Protection Agency
Office of Press Service
Waterside Mall West Tower
Room 329
Washington, DC 20460
  (202) 755-0344
    Director: Marlin Fitzwater

U.S.
Environmental Protection Agency
Office of Public Awareness
Waterside Mall West Tower
Room 311C
Washington, DC 20460

(202) 755-0700
Director: Joan Martin Nicholson

U.S.
Environmental Protection Agency
Office of Public Awareness
Information Development
Waterside Mall West Tower
Room 239
Washington, DC 20460
   (202) 755-0872
   Coordinator: Alex MacDonald

U.S.
Environmental Protection Agency
Office of Regional and Intergovern-
   mental Operations
Waterside Mall West Tower
Room 1137
Washington, DC 20460
   (202) 755-0444
   Acting Director: K. Kirke Harper

U.S.
Environmental Protection Agency
Office of the Administrator
Waterside Mall West Tower
Room 1200
Washington, DC 20460
   (202) 755-2700
   Administrator: Douglas Costle

U.S.
Environmental Protection Agency
Office of the Administrator
Office of EPA Land Use Coordination
Waterside Mall West Tower
Room 1224
Washington, DC 20460
   (202) 755-0442
   Director: John Gustafson

U.S.
Environmental Protection Agency
Office of the Administrator
Public Participation
Waterside Mall West Tower
Room 1225
Washington, DC 20460
   (202) 755-0425
   Sharon Francis

U.S.
Environmental Protection Agency
Public Information Center
401 M Street, S.W.
Washington, DC 20460
   (202) 755-0707
   Director

U.S.
Environmental Protection Agency
Region I
Room 2203
Kennedy Federal Building
Boston, MA 02203
   (617) 223-7210
   Administrator: William Adams, Jr.

U.S.
Environmental Protection Agency
Region II
Room 1009
26 Federal Plaza
New York, NY 10007
   (212) 264-2525
   Administrator: Eckhardt Beck

U.S.
Environmental Protection Agency
Region III
Curtis Building
6th and Walnut Streets
Philadelphia, PA 19106
   (215) 597-9814
   Administrator: Jack Schramm

U.S.
Environmental Protection Agency
Region IV
345 Courtland Street, N.E.
Atlanta, GA 30308
   (404) 881-4727
   Administrator: Rebecca Hanmer

U.S.
Environmental Protection Agency
Region IV
EIS Branch
345 Courtland Street
Atlanta, GA 30308
   (404) 881-4727
   Chief: John Hagan, III

U.S.
Environmental Protection Agency
Region IV
EIS Review Section
345 Courtland Street
Atlanta, GA   30308
   (404) 881-4727
   Section Chief: Sheppard Moore

U.S.
Environmental Protection Agency
Region IV
EIS Preparation Section
345 Courtland Street
Atlanta, GA   30308
   (404) 881-4727
   Section Chief: Robert Howard

U.S.
Environmental Protection Agency
Region V
230 South Dearborn Street
Chicago, IL   60604
   (312) 353-2000
   Administrator: John McGuire

U.S.
Environmental Protection Agency
Region V
Environmental Impact Review Staff
230 Dearborn Street
Chicago, IL   60604
   (312) 353-2000
   Section Chief: Barbara Taylor

U.S.
Environmental Protection Agency
Region VI
First International Building
1201 Elm Street
Dallas, TX   75270
   (214) 767-2716
   Regional EIS Coordinator: Clinton
      Spotts

U.S.
Environmental Protection Agency
Region VII
324 East 11th Street
Kansas City, MO   64108
   (816) 374-5493
   EIS Coordinator: Edward Vest

U.S.
Environmental Protection Agency
Region VIII
Room 900
Lincoln Tower Building
1860 Lincoln Street
Denver, CO   80203
   (303) 837-3895`
   Administrator: Alan Merson

U.S.
Environmental Protection Agency
Region IX
215 Fremont Street
San Francisco, CA   94105
   (415) 556-2320
   Administrator: Paul DeFalco, Jr.

U.S.
Environmental Protection Agency
Region X
1200 Sixth Avenue
Seattle, WA   98101
   (206) 442-1220
   Administrator: Donald Dubos

U.S.
Environmental Protection Agency
Region X
EIS Review Team
1200 Sixth Avenue
Seattle, WA   98101
   (206) 442-1285
   Team Leader: Dan Steinborn

U.S.
Environmental Protection Agency
Region X
Environmental Evaluation Branch
1200 Sixth Avenue
Seattle, WA   98101
   (206) 442-1285
   Branch Chief: Lisa Corbyn

U.S.
Environmental Protection Agency
Region X
Environmental Impact Statement
   Preparation Team
1200 Sixth Avenue
Seattle, WA   98101
   (206) 442-4011

Team Leader: Roger Mochnick

U.S.
Federal Communications Commission
Office of General Counsel
Washington, DC 20554
  (202) 632-6393
  Coordinator: Upton Guthery

U.S.
Federal Energy Regulatory Commission
Environmental Analysis Branch
Washington, DC 20426
  (202) 376-1768
  Chief: Quentin Edson

U.S.
Federal Energy Regulatory Commission
Environmental Analysis Branch
Conservation Section
Washington, DC 20426
  (202) 376-1909
  Chief: Dean Shumway

U.S.
Federal Energy Regulatory Commission
Environmental Analysis Branch
Environmental Studies Section
Washington, DC 20426
  (202) 376-4336
  Chief: Clarence Blackstock

U.S.
Federal Energy Regulatory Commission
Environmental Analysis Branch
Recreation and Land Use Section
Washington, DC 20426
  (202) 376-4312
  Chief: John Young, Jr.

U.S.
Federal Maritime Commission
Office of Environmental Analysis
1100 L Street, N.W.
Washington, DC 20573
  (202) 523-5835
  Chief: Paul Gonzalez

U.S.
General Services Administration
18th and F Streets, N.W.
Washington, DC 20450

  (202) 655-4000
  Administrator: Paul Goulding

U.S.
General Services Administration
Eastern Division
18th and F Streets, N.W.
Washington, DC 20405
  (202) 472-1082
  Director: Norman Berky

U.S.
General Services Administration
Environmental Affairs Division
18th and F Streets, N.W.
Washington, DC 20405
  (202) 566-1416
  Acting Director: Carl Penland

U.S.
General Services Administration
National Capital Region
7th and D Streets, N.W.
Washington, DC 20407
  (202) 655-4000
  Administrator: Walter Kallaur

U.S.
General Services Administration
Region I
McCormick Post Office and Courthouse
Boston, MA 02109
  (617) 223-2601
  Administrator: Lawrence Bretta

U.S.
General Services Administration
Region II
26 Federal Plaza
New York, NY 10007
  (212) 264-2600
  Administrator: Gerald Turetsky

U.S.
General Services Administration
Region III
9th and Market Streets
Philadelphia, PA 19107
  (202) 472-1100
  Administrator: Clarence Lee, Jr.

U.S.

General Services Administration
Region IV
1776 Peachtree Street, N.W.
Atlanta, GA   30309
  (404) 881-4600
  Administrator: Wesley Johnson

U.S.
General Services Administration
Region V
230 South Dearborn Street
Chicago, IL   60604
  (312) 353-5395
  Administrator: Clarence Sochowski

U.S.
General Services Administration
Region VI
1500 East Banister Road
Kansas City, MO· 64131
  (816) 926-7201
  Administrator: Hazen Harvell

U.S.
General Services Administration
Region VII
819 Taylor Street
Fort Worth, TX   76102
  (817) 334-2321
  Administrator: Ann Doughty

U.S.
General Services Administration
Region VIII
Denver Federal Center
Denver, CO   80225
  (303) 234-4171
  Administrator: Dennis Jensen

U.S.
General Services Administration
Region IX
525 Market Street
San Francisco, CA   94105
  (415) 556-3221
  Administrator: Joseph Williams

U.S.
General Services Administration
Region X
GSA Center
Auburn, WA   98002

  (206) 883-6500
  Administrator: Richard Casad

U.S.
General Services Administration
Western Division
18th and F Streets, N.W.
Washington, DC   20405
  (202) 472-1082
  Director: Earl Jones

U.S.
Great Lakes Basin Commission
P.O. Box 999
3475 Plymouth Road
Ann Arbor, MI   48106
  (313) 668-2300
  Chairman: Lee Botts

U.S.
International Boundry and Water
  Commission
United States Section
4110 Rio Bravo
El Paso, TX   79902
  (915) 543-7313
  Chief, Planning and Reports Branch:
    John Vandertulip

U.S.
Interstate Commerce Commission
Section of Energy and Environment
Room 3371
12th and Constitution Avenue, N.W.
Washington, DC   20423
  (202) 275-7658
  Chief: Carl Bausch

U.S.
Missouri River Basin Commission
10050 Regency Circle
Suite 403
Omaha, NE   68114
  (402) 864-9351
  Chairman: John Acord

U.S.
National Aeronautics and Space Admin-
  istration
Management and Support Office (LB-4)
400 Maryland Avenue, S.W.
Washington, DC   20546

(202) 755-8384
Director: Nathaniel Cohen

U.S.
National Capitol Planning Commission
Office of Environmental Affairs
Washington, DC 20576
(202) 724-0200
Chief: Patricia Crawford

U.S.
National Endowment for the Arts
Architecture and Environmental Arts
  Program
Washington, DC 20506
(202) 634-4276
Director

U.S.
National Science Foundation
Astronomical, Atmospheric Earth
  and Ocean Sciences
Room 641
1800 G Street, N.W.
Washington, DC 20550
(202) 632-7360
  Assistant Director: Dr. Francis
    Johnson

U.S.
National Science Foundation
Science Resources Studies Division
1800 G Street, N.W.
Washington, DC 20550
(202) 634-4625
Director

U.S.
New England River Basins Commission
53 State Street
Boston, MA 02109
(617) 223-6244
Chairman: John Ehvenfeld

U.S.
Nuclear Regulatory Commission
Division of Site Safety and Environ-
  mental Analysis
Cost-Benefit Analysis Branch
Washington, DC 20555
(301) 492-7161
Chief: B. Joe Youngblood

U.S.
Ohio River Basins Commission
36 East 4th Street
Suite 208-220
Cincinnati, OH 45202
(513) 684-3831
Chairman: Fred Morr

U.S.
Pacific Northwest River Basins
  Commission
P.O. Box 908
Vancouver, WA 98660
(206) 422-9307
Chairman: Melvin Gordon

U.S.
Postal Service
Real Estate and Building Department
Office of Program Planning
Washington, DC 20260
(202) 245-4304
Director: Robert Coven

U.S.
Susquehanna River Basin Commission
18th and C Streets, N.W.
Washington, DC 20240
(202) 343-4091
Commissioner: Patrick Delaney

U.S.
Tennessee Valley Authority
Environmental Planning Division
264-401 Building
Chattanooga, TN 37401
(615) 755-3155
H.R. Hickey

U.S.
Tennessee Valley Authority
Environmental Research and Develop-
  ment
400 Commerce Avenue
Knoxville, TN 37902
(615) 632-2101
Director: Harry Moore

U.S.
Tennessee Valley Authority
Office of Environmental Quality
Forestry Building

Norris, TN 37828
  (615) 632-6450
  Director: Dr. Mohamed T. El-Ashry

U.S.
Upper Mississippi River Basin
  Commission
Federal Building
Room 510
Fort Snelling
Twin Cities, MN 55111
  (612) 725-4690
  Executive Director: William Walton

U.S.
Veterans Administration
Office of Environmental Activities
  (004A)
810 Vermont Avenue
Washington, DC 20420
  (202) 389-2526
  Director: Willard Sitler

Upper Minnesota Valley Regional
  Development Commission
323 West Schlieman Avenue
Appleton, MN 56208
  (612) 289-1981
  Director

Upper Mississippi River Conservation
  Committee
1830 Second Avenue
Rock Island, IL 61201
  (309) 788-3991
  Chairman: James Addis

Urban Land Institute
1200-18th Street, N.W.
Washington, DC 20036
  (202) 331-8500
  Director

Utah Association of Counties
10 West Broadway
Suite 311
Salt Lake City, UT 84101
  (801) 364-3583
  Executive Director: Jack Tanner

Utah
Department of Natural Resources

400 Empire Building
231 East Fourth South
Salt Lake City, UT 84111
  (801) 533-5356
  Executive Director: Gordon Harmston

Utah
Department of Natural Resources
Central Region
176 East Center Street
Provo, UT 84601
  (801) 373-4774
  Regional Supervisor: LaVar Ware

Utah
Department of Community and Economic
  Development
Environmental Coordinating Committee
124 State Capitol
Salt Lake City, UT 84111
  (801) 533-5794
  Ken Crovins

Utah
Department of Natural Resources
Division of State Lands
411 Empire Building
231 East Fourth South
Salt Lake City, UT 84111
  (801) 533-5381
  Director: William Dinehart

Utah
Department of Natural Resources
Northeastern Region
671 West 100 North
Vernal, UT 84078
  (801) 789-3103
  Regional Supervisor: Donald Smith

Utah
Department of Natural Resources
Northern Region
166 East 4600 South
Ogden, UT 84403
  (801) 392-6001
  Regional Supervisor: Jack Rensel

Utah
Department of Natural Resources
Southeastern Region
455 West Railroad Avenue

Price, UT 84501
  (801) 637-3310
  Regional Supervisor: Larry Wilson

Utah
Department of Natural Resources
Southern Region
622 North Main Street
Box 606
Cedar City, UT 84720
  (801) 586-6803
  Regional Supervisor: Clair Jensen

Utah
Health Service Branch
150 W.N. Temple
Room 426
Salt Lake City, UT 84110
  (801) 533-6121
  James Clise

Utah
League of Cities and Towns
10 West Broadway
Suite 304-307
Salt Lake City, UT 84101
  (801) 328-1601
  Executive Director: Bennie Schmiett

Utah Wildlife and Outdoor Recreation
  Federation
P.O. Box 15636
Salt Lake City, UT 84115
  (801) 521-4567
  President: Steven Wiseman

Utica City
Department of Urban and Economic
  Development
City Hall
1 Kennedy Plaza
Utica, NY 13502
  (315) 798-3236
  Senior Planner: Edward Gorski

Vallejo City
Planning Department
City Hall
Vallejo, CA 94590
  (707) 553-4326
  Director: Don Patterson

Valley County
Planning Department
County Courthouse
Glasgow, MT 59230
  (406) 228-4613
  Director

Valley Regional Planning Agency Valley
  Council of Governments
Derby Train Station
Main Street
Derby, CT 06418
  (203) 735-8689
  Director

Vance County
Planning Department
County Courthouse
Henderson, NC 27536
  (919) 492-0031
  Director

Vanderburgh County
Planning Department
County Courthouse
Evansville, IN 47708
  (812) 426-5160
  Director

Ventura County
Resource Management Agency
Planning Department
800 South Victoria Avenue
Ventura, CA 93009
  (805) 654-2267
  EIS Contact: Robert Laughlin

Vermillion County
Area Plan Commission
County Courthouse
Newport, IN 47966
  (317) 492-3570
  Director

Vermont
Agency of Environmental Conservation
Environmental Board
Montpelier, VT 05602
  (802) 828-3309
  Executive Officer: Richard Cowart

Vermont

Agency of Environmental Conservation
Environmental Conservation Planning
Montpelier, VT 05602
   (802) 828-3357
   Director: Edward Koenemann

Vermont
Agency of Environmental Conservation
Environmental Protection Division
Montpelier, VT 05602
   (802) 828-3341
   Canute Dalmasse

Vermont Association of Conservation
   Districts
Rochester, VT 05767
   (802) 767-3970
   President: Merle Severy

Vermont Institute of Natural Science
Church Hill Road
Woodstock, VT 05091
   (802) 457-2779
   President: Willis Curtis

Vermont League of Cities and Towns
118 Main Street
Montpelier, VT 05602
   (802) 229-9111
   Executive Director: Robert Stewart

Vermont Natural Resources Council
26 State Street
Montpelier, VT 05602
   (802) 223-2328
   President: David Marvin

Vermont Tomorrow
5 State Street
Montpelier, VT 05602
   (802) 223-6067
   Executive Director: David Goldberg

Vernon County
Planning Department
County Courthouse
Viroqua, WI 54665
   (608) 637-3569
   Director

Vigo County
Planning Department

County Courthouse
Terre Haute, IN 47801
   (812) 234-2671
   Director

Vilas County
Zoning, Planning and Pollution Control
P.O. Box 340
Courthouse
Eagle River, WI 54521
   (715) 479-8902
   Director

Virginia Association of Counties
P.O. Box 6306
Charlottesville, VA 22906
   (804) 973-7557
   Executive Director: George Long

Virginia Association of Soil and Water
   Conservation Districts
Suite 800
830 East Main Street
Richmond, VA 23219
   (804) 786-2064
   President: J. Royall Robertson

Virginia Beach City
Planning Department
Municipal Center
Virginia Beach, VA 23456
   (804) 427-4111
   EIS Contact: William Whitney

Virginia
Council on the Environment
903 Ninth Street
Richmond, VA 23219
   (804) 786-4500
   EIS Coordinator: Charles Ellis, III

Virginia
Department of Conservation and
   Economic Development
1100 State Office Building
Richmond, VA 23219
   (804) 786-2121
   Director: Fred Walker

Virginia
Marine Resources Commission
P.O. Box 756

Newport News, VA   23607
  (804) 245-2811
  Assistant Commissioner for Environ-
    mental Affairs: Norman Larsen

Virginia Municipal League
311 Ironfronts
1011 East Main Street
Richmond, VA   23206
  (804) 649-8471
  Executive Director: Richard DeClair

Virginia Planning District Commission
145 West Campbell Avenue
Roanoke, VA   24010
  (703) 343-4417
  Director

Virginia Soil and Water Conservation
  Commission
Suite 800
830 East Main Street
Richmond, VA   23219
  (804) 786-2064
  Director: Joseph Willson, Jr.

Virgin Islands Conservation District
P.O. Box 2144
St. Thomas, VI   00801
  (809) 774-0428
  Chairman: Clifford Benjamin

Virgin Islands Conservation Society
P.O. Box 4187
St. Thomas, VI   00801
  (809) 775-3225
  President: Dr. Edward Towle

Virgin Islands
Department of Conservation and
  Cultural Affairs
P.O. Box 4340
St. Thomas, VI   00801
  (809) 774-3320
  Commissioner: Virdin Brown

Wabasha County
Planning Department
County Courthouse
Wabasha, MN   55981
  (612) 565-3978
  Director

Wabash County
Planning Deparment
County Courthouse
Wabash, IN   46992
  (219) 563-5217
  Director

Wadena County
Zoning Office
County Courthouse
Wadena, MN   56482
  (218) 631-2852
  EIS Contact: G.J. Kempf

Wahkiakum County
Planning Department
County Courthouse
Cathlamet, WA   98612
  (206) 795-3219
  Director

Wake County
Planning Department
P.O. Box 550
Raleigh, NC   27602
  (919) 755-6160
  Director: John Scott

Walla Walla County
Planning Department
P.O. Box 478
Walla Walla, WA   99362
  (509) 525-6161
  Associate Planner: Peter Skowlund

Walla Walla Regional Planning
P.O. Box 905
Walla Walla, WA   99362
  (509) 529-8260
  Director: Jim Beard

Waltham City
Planning Department
City Hall
Waltham, MA   02154
  (617) 893-4040
  Director

Walworth County
Planning Department
County Courthouse
Selby, SD   57472

(605) 649-7311
Director

Walworth County
Planning Department
County Courthouse
Elkhorn, WI 53121
   (414) 723-4900
   Director

Warren County
Planning Department
County Courthouse
Williamsport, IN 47993
   (317) 762-3275
   Director

Warren County
Planning Department
County Courthouse
Lake George, NY 12845
   (518) 792-9951
   Director: J. Bryan Harrison

Warren County
County Manager's Office
P.O. Box 619
Warrenton, NC 27589
   (919) 257-3261
   EIS Contact: Glenwood Newsome

Warren County
Planning Department
P.O. Box 908
Fort Royal, VA 22630
   (703) 636-9973
   Director

Warrick County
Planning Department
County Courthouse
Boonville, IN 46701
   (812) 897-3590
   Director

Waseca County
Office of Zoning and Planning
County Courthouse
Waseca, MN 56093
   (507) 835-5600
   EIS Contact: David Zimmerman

Washburn County
Planning Department
County Courthouse
Shell Lake, WI 54871
   (715) 468-7808
   Director

Washington Association of Cities
4719 Brooklyn Avenue, N.E.
Seattle, WA 98105
   (206) 543-9050
   Executive Director: Kent Swisher

Washington Association of Conservation
   Districts
P.O. Box 4128
Tumwater, WA 98501
   (206) 352-1388
   President: Charles O'Neill, Jr.

Washington Association of County
   Officials
105 East Eighth Avenue
Suite 307
Olympia, WA 98501
   (206) 943-1812
   Executive Director: Fred Saeger

Washington County
Planning Department
County Courthouse
Salem, IN 47167
   (812) 883-5748
   Director

Washington County
Planning Department
County Courthouse
Hagerstown, MD 21740
   (301) 791-3090
   Director

Washington County
Planning Department
County Courthouse
Stillwater, MN 55082
   (612) 439-3220
   Director

Washington County
Planning Department
County Courthouse

Washington County
Planning Department
County Courthouse
Fort Edward, NY 12828
 (518) 747-3374
 Director

Washington County
Planning Department
County Courthouse
Plymouth, NC 27962
 (919) 793-2631
 Director

Washington County
Planning Department
203 Court Street
Abingdon, VA 24210
 (703) 628-2983
 County Administrator: Robinson
  Worth, Jr.

Washington County
Planning Department
County Courthouse
West Bend, WI 53095
 (414) 334-3491
 Director

Washington
Department of Ecology
Olympia, WA 98504
 (206) 753-2800
 Director

Washington
Department of Ecology
Environmental Review Section
Olympia, WA 98504
 (206) 753-2800
 Section Head: Greg Sorlie

Washington
Department of Ecology
Shorelands Division
Olympia, WA 98504
 (206) 753-2800
 Division Supervisor: Duane Wegner

Washington
Department of Natural Resources
Public Lands Building

Olympia, WA 98504
 (206) 753-5327
 Administrator: Bert Cole

Washington Ecology Center
901 Larch Avenue
Takoma Park, MD 20012
 (301) 270-3168
 Executive Director: David Paris

Washington Environmental Council
107 South Main
Seattle, WA 98104
 (206) 623-1483
 President: Helen Engle

Washington State Association of
 Counties
6730 Martin Way, N.E.
Olympia, WA 98506
 (206) 491-7100
 Executive Director: Jack Rogers

Washington State Conservation
 Commission
PV-11
Olympia, WA 98504
 (206) 753-3895
 Chairman: William Schmidtman

Washington State Sportsmen's Council
P.O. Box 98236
Tacoma, WA 98499
 (206) 564-6840
 President: John Stone

Watauga County
Planning Department
County Courthouse
Boone, NC 28607
 (704) 264-1300
 Director

Waterbury City
Planning Department
City Hall
Waterbury, CT 06702
 (203) 756-9494
 Director

Watonwan County
Planning Department

County Courthouse
St. James, MN 56081
  (507) 375-3541
  Director

Waukesha County
Planning Department
County Courthouse
Waukesha, WI 53705
  (414) 544-8227
  Director

Waupaca County
Planning Department
County Courthouse
Waupaca, WI 54981
  (715) 258-2128
  Director

Waushara County
Planning Department
County Courthouse
Wautoma, WI 54982
  (414) 787-2320
  Director

Wauwatosa City
Planning Department
City Hall
Wauwatosa, WI 53213
  (414) 258-3000
  Director

Wayne County
Planning Department
County Courthouse
Richmond, IN 47374
  (317) 966-7541
  Director

Wayne County
Planning Department
County Courthouse
Lyons, NY 14489
  (315) 946-9767
  Director

Wayne County
Planning Department
County Courthouse
Goldsboro, NC 27530
  (919) 735-4331

Director

Wells County
Area Plan Commission
County Courthouse
Bluffton, IN 46714
  (219) 824-0183
  Director: Paul Dotterer

West Allis City
Planning Department
City Hall
West Allis, WI 53214
  (414) 476-4340
  Director

West Central Indiana Economic Development District
700 Wabash Avenue
P.O. Box 359
Terre Haute, IN 47808
  (812) 238-1561
  Director

West Central Regional Development
  Commission
Fergus Falls Community College
Fergus Falls, MN 56537
  (218) 739-3356
  Director

West Central Wisconsin Regional
  Planning Commission
124½ Graham Avenue
Eau Claire, WI 54701
  (715) 836-2918
  Executive Director: John Lohrentz

Westchester County
Planning Department
County Office Building
White Plains, NY 10601
  (914) 682-2000
  Director

West Covina City
Planning Department
City Hall
West Covina, CA 91790
  (213) 962-8631
  Director

Western Pennsylvania Conservancy
316 Fourth Avenue
Pittsburgh, PA 15222
    (412) 288-2777
    Chairman: Joshua Whetzel

Western Piedmont Council of Govern-
    ments
30 Third Street, N.W.
Old City Hall
Hickory, NC 28601
    (704) 328-2936
    Director

Western Regional Environmental
    Education Council
P.O. Box 3503
Portland, OR 97208
    (503) 229-5551
    Chairman: Cliff Hamilton

West Hartford Town
Community Development Program
Town Hall
West Hartford, CT 06107
    (203) 236-3231
    EIS Contact: Kerron Barnes

West Haven City
Planning Department
City Hall
West Haven, CT 06516
    (203) 934-3421
    Director

West Michigan Environmental Action
    Council
1324 Lake Drive, S.E.
Grand Rapids, MI 49506
    (616) 451-3051
    Executive Director: Roger Conner

Westminster City
Planning Department
8200 Westminster Avenue
Westminster, CA 92683
    (714) 898-3311
    Director

Westmoreland County
Land Use Administration
P.O. Box 467

Montross, VA 22520
    (804) 493-8911
    Director

West Piedmont Planning District
    Commission
18-22 Walnut Street
P.O. Box 1191
Martinsville, VA 24112
    (703) 638-3987
    Director

West Virginia Association of County
    Officials
1018 Kanawha Boulevard East
Suite 207
Charleston, WV 25301
    (304) 346-0592
    Executive Director: Gene Elkins

West Virginia
Department of Natural Resources
1800 Washington Street, East
Charleston, WV 23505
    (304) 348-2754
    Director: David Callaghan

West Virginia Municipal League
1615 Washington Street, East
Charleston, WV 25311
    (304) 343-9201
    Executive Director: Betty Dean

West Virginia Soil and Water Conser-
    vation District Supervisors Assoc-
    iation
P.O. Box 429
Lewisburg, WV 24901
    (304) 645-2309
    President: David Tuckwiller

Weymouth Town
Planning Board
Town Hall
75 Middle Street
Weymouth, MA 02189
    (617) 335-2000
    Senior Planner: Roderick Fuqua

Whatcom County
Planning Department
County Courthouse

Cathlamet, WA 98612
  (206) 676-6700
  EIS Contact: Robert Bacon

Whatcom County Council of Governments
625 Cornwall One
Bellingham, WA 98225
  (206) 676-2503
  Director

Wheatland County
Planning Department
County Courthouse
Harlowton, MT 59036
  (406) 632-5621
  Director

White County
Planning Department
Box 851
Monticello, IN 47960
  (219) 583-7032
  Executive Director: C.E. Krecek

White Plains City
Planning Department
City Hall
White Plains, NY 10602
  (914) 949-4800
  Director

Whitley County
Planning Department
County Courthouse
Columbia City, IN 46725
  (219) 248-8212
  Director

Whitman County
Planning Department
County Courthouse
Colfax, WA 99111
  (509) 397-2522
  Director

Whitman County Regional Planning
  Council
Old National Bank Building
Room 8
Colfax, WA 99111
  (509) 397-4303
  Director

Whittier City
Planning Department
City Hall
Whittier, CA 90602
  (213) 698-2551
  Director

Wibaux County
Planning Department
County Courthouse
Wibaux, MT 59353
  (406) 795-2433
  Director

Wicomico County
Planning Department
P.O. Box 870
Salisbury, MD 21801
  (301) 749-5127
  Director

Wilderness Society
1901 Pennsylvania Avenue, N.W.
Washington, DC 20006
  (202) 293-2732
  President: Charles Stoddard

Wilkes County
Planning Department
County Courthouse
Wilkesboro, NC 28697
  (919) 667-5111
  Director

Wilkin County
Planning Department
County Courthouse
Breckenridge, MN 56520
  (218) 643-4972
  Director

Wilson County
Planning Department
P.O. Box 1228
Wilson, NC 27893
  (919) 237-3913
  Director: Garry Mercer

Windham Regional Planning Agency
21 Church Street
Williamantic, CT 06226
  (203) 456-2221

Director

Winnebago County
Planning Department
County Courthouse
Oshkosh, WI 54901
   (414) 235-2500
   Director

Winona County
Planning Department
County Courthouse
Winona, MN 55987
   (507) 452-3337
   Director

Winston-Salem City
Planning Board
P.O. Box 2511
Winston-Salem, NC 27102
   (919) 727-2123
   EIS Contact: James Yarbrough

Wisconsin Association of Soil and
   Water Conservation Districts
Rt. 1
Portage, WI 53901
   (608) 742-3449
   President: Frank Kiefer

Wisconsin County Boards Association
122 West Washington Avenue
Suite 200
Madison, WI 53703
   (608) 266-6480
   Executive Director: Robert
      Mortensen

Wisconsin
Department of Natural Resources
Box 7921
Madison, WI 53707
   (608) 266-2621
   Secretary: A.S. Earl

Wisconsin
Department of Natural Resources
Bureau of Environmental Impact
Box 7921
Madison, WI 53707
   (608) 266-0860
   Director: Stan Druckenmiller

Wisconsin
Department of Natural Resources
Division of Environmental Standards
Box 7921
Madison, WI 53707
   (608) 266-1099
   Administrator: Thomas Kroehn

Wisconsin Environmental Decade
114 East Miffin Street
Madison, WI 53703
   (608) 251-7020
   Director: John Neess

Wisconsin League of Municipalities
122 West Washington Avenue
Madison, WI 53703
   (608) 255-7291
   Executive Director: Ed Johnson

Wisconsin State Board of Soil and
   Water Conservation Districts
1815 University Avenue
Madison, WI 53706
   (608) 262-2634
   Executive Secretary: Eugene Savage

Wise County
Planning Department
P.O. Box 570
Wise, VA 24293
   (703) 328-6111
   Director

Wood County
Planning Department
County Courthouse
400 Market Street
P.O. Box 825
Wisconsin Rapids, WI 54494
   (715) 423-3000
   EIS Contact: Marvin Krzykowski

Worcester County
Planning Department
County Courthouse
Snow Hill, MD 21863
   (301) 632-1194
   Director

Worcester City
Planning Department

City Hall
Worcester, MA 01608
  (617) 798-8151
  Director

Worcester County
Engineering Department
County Courthouse
Room 101
Worcester, MA 01608
  (617) 798-7706
  County Engineer: John O'Toole

Wright County
Planning Department
County Courthouse
Buffalo, MN 55313
  (612) 682-3900
  Director

Wyoming Association of Conservation
  Districts
Box 366
Burlington, WY 82411
  (307) 762-3443
  President: Stanley Preator

Wyoming Association of Municipalities
P.O. Box 2535
Cheyenne, WY 82001
  (307) 632-0398
  Executive Director: Robert Contine

Wyoming County
Planning Department
County Courthouse
Warsaw, NY 14569
  (716) 786-2264
  Director

Wyoming County Commissioners
  Association
P.O. Box 86
Cheyenne, WY 82001
  (307) 632-5409
  Executive Director: Norm Cable

Wyoming
Environmental Quality Department
Hathaway Building
Cheyenne, WY 82002
  (307) 777-7391

Director: Robert Sundin

Wyoming Outdoor Council
2003 Central
P.O. Box 1184
Cheyenne, WY 82001
  (307) 635-3416
  President: Leslie Petersen

Wythe County
Planning Department
County Courthouse
Wytheville, VA 24382
  (703) 228-4991
  Director

Yadkin County
Planning Department
P.O. Box 146
Yadkinville, NC 27055
  (919) 679-8838
  Director

Yakima County
Planning Department
County Courthouse
Yakima, WA 98901
  (509) 575-4111
  Director

Yakima County Conference of Govern-
  ments
County Courthouse
Yakima, WA 98901
  (509) 575-4124
  Director

Yancey County
Planning Department
County Courthouse
Burnsville, NC 28714
  (704) 682-3971
  Director

Yankton County
Planning Department
P.O. Box 137
Yankton, SD 57078
  (605) 665-2143
  Director

Yates County

Planning Department
County Building Annex
431 Liberty Street
Penn Yan, NY 14527
  (315) 536-2531
  EIS Contact: Bob Napoli

Yellow Medicine County
Planning Department
County Courthouse
Granite Falls, MN 56241
  (612) 564-3132
  Director

Yellowstone County
Planning Department
County Courthouse
Billings, MT 59101
  (406) 252-5181
  Director

Yolo County
Planning Department
292 West Beamer Street
Woodland, CA 95695
  (916) 666-8204
  EIS Contact: Doris Michaels

Yonkers City
Planning Department
City Hall
Yonkers, NY 10701
  (914) 963-3980
  Director

York County
Planning Department
P.O. Box 532
Yorktown, VA 23690
  (804) 898-3434
  Director

Yuba County
Planning Department
County Courthouse
Marysville, CA 95901
  (916) 674-6341
  Director

Ziebach County
Planning Department
County Courthouse

Dupree, SD 57623
  (605) 365-5159
  Director

2. LEGAL-RELATED AGENCIES AND ORGANIZATIONS

Alameda County Bar Association
c/o Harold Norton
405-14th Street
Suite 208
Oakland, CA 94612
(415) 893-7160

Albany County Bar Association
c/o Jeanne Bennison
Albany County Court House
Albany, NY 12207
(518) 445-7691

Alexandria Bar Association
c/o William O'Neil
307 North Washington Street
Alexandria, VA 22314
(703) 836-5757

Allen County Bar Association
c/o Barbara Carto
1904 Wayne Bank Building
Fort Wayne, IN 46802
(219) 423-2358

American Bar Association
Standing Committee on Environmental
  Law
1800 M Street, N.W.
Washington, DC 20036
(202) 331-2278
Chairman: Carroll Gilliam

Baltimore City Bar Association
c/o Robert Ashman

629 Civil Courts Building
Baltimore, MD 21202
(301) 539-5936

Baltimore County Bar Association
c/o Lee Stuart Thompson
800 Equitable Towson Building
Towson, MD 21204
(301) 821-6600

Bar Association of Nassau County
c/o William Jackson, Jr.
15 and West Streets
Mineola, NY 11501
(516) 747-4070

Beverly Hills Bar Association
c/o Laura Aatlo
300 South Beverly Drive
210
Beverly Hills, CA 90212
(213) 553-6644

Boston Bar Association
c/o Frederick Norton, Jr.
16 Beacon Street
Boston, MA 02108
(617) 742-0615

Bridgeport Bar Association
c/o Harold Pitt
955 Main Street
Bridgeport, CT 06604
(203) 334-6539

Bronx County Bar Association
c/o Esther Kerrigan
851 Grand Concourse
Bronx, NY 10451
    (212) 293-5600

Brooklyn Bar Association
c/o Frederick Gross
123 Remsen Street
Brooklyn, NY 11201
    (212) 624-0675

Broome County Bar Association
c/o Sandra Hian
P.O. Box 1766
Binghamton, NY 13901
    (607) 772-2196

California State Board
c/o Richard Morris
555 Franklin Street
San Francisco, CA 94102
    (415) 561-8300

Connecticut Bar Association
c/o Daniel Hovey
15 Lewis Street
Hartford, CT 06103
    (203) 249-9141

Conservation Law Foundation of
    New England
3 Joy Street
Boston, MA 02108
    (617) 742-2540
    Executive Director: Douglas Foy

Dane County Bar Association
c/o Dorothy Heil
Madison, WI 53701
    (608) 256-5617

Environmental Defense Fund
475 Park Avenue South
New York, NY 10016
    (212) 686-4191
    Executive Director: Arlie Schardt

Environmental Law Institute
1346 Connecticut Avenue, N.W.
Suite 600
Washington, DC 20036

    (202) 452-9600
    President: Frederick Anderson

Erie County Bar Association
c/o Carol Scal
1758 Statler Hilton Hotel
Buffalo, NY 14203
    (716) 852-8687

Fairfax Bar Association
c/o Rosalie Small
4000 Chain Bridge Road
Fairfax, VA 22030
    (703) 273-6860

Fresno County Bar Association
c/o Val Weston
409 T.W. Patterson Building
Fresno, CA 93721
    (209) 264-0137

Hampden County Bar Association
c/o Judith Potter
50 State Street
Springfield, MA 01103
    (413) 732-4648

Hartford County Bar Association
c/o Mary St. Clair
266 Pearl Street
Hartford, CT 06103
    (203) 525-8106

Hawaii State Bar
c/o Elenor Pierce
P.O. Box 26
Honolulu, HI 96810
    (808) 537-1868

Hennepin County Bar Association
c/o George Claseman
700 Cargill Building
Minneapolis, MN 55402
    (612) 335-0921

Indianapolis Bar Association
c/o Rosalie Felton
One Indiana Square
Suite 2550
Indianapolis, IN 46204
    (317) 632-8240

Indiana State Bar Association
c/o E.B. Lyle
230 East Ohio Street
Indianapolis, IN 46204
  (317) 639-5465

Long Beach Bar Association
c/o Nila Alcock
444 West Ocean Boulevard
Suite 500
Long Beach, CA 90802
  (213) 432-5913

Los Angeles County Bar Association
c/o David Ellwanger
606 South Oliver Street
Suite 1212
Los Angeles, CA 90014
  (213) 622-8682

Marin County Bar Association
c/o Jeanette Stewart
1010 B Street
Suite 419
San Rafael, CA 94901
  (415) 453-5505

Maryland State Bar Association
c/o William Smith, Jr.
905 Keyser Building
Baltimore, MD 21202
  (301) 685-7878

Massachusetts Bar Association
c/o Carl Modecki
One Center Plaza
Boston, MA 02108
  (617) 523-4529

Middlesex County Bar Association
c/o William Highgas Court House
Cambridge, MA 02141
  (617) 494-4150

Milwaukee Bar Association
c/o Georgeanna Rude
610 North Jackson Street
Milwaukee, WI 53203
  (414) 271-3833

Minnesota State Bar Association
c/o Gerald Regnier

100 Minnesota Federal Building
Minneapolis, MN 55402
  (612) 335-1183

Monroe County Bar Association
c/o Milford Wheeler
1125 First Federal Plaza
Rochester, NY 14614
  (716) 546-1817

Montana State Bar
c/o Ken Parcell
P.O. Box 4669
Helena, MT 59601
  (406) 442-7660

Montgomery County Bar Association
c/o Willima Huffman
17 West Jefferson Street
Rockville, MD 20850
  (301) 424-3454

Mt. Diablo Bar Association
c/o Lillian Galvin
1910 Olympic Boulevard
Suite 325
Walnut Creek, CA
  (415) 933-5650

New Haven County Bar Association
c/o Susan Shimelmann
205 Church Street
New Haven, CT 06509
  (203) 562-5750

New York Association of the Bar
c/o Paul DeWitt
42 West 44th Street
New York, NY 10036
  (212) 840-3550

New York County Lawyers Association
c/o William Greene, III
14 Vesey Street
New York, NY 10007
  (212) 267-6646

New York State Bar Association
c/o William Carroll
One Elk Street
Albany, NY 12207
  (518) 445-1211

New York Womens Bar Association
c/o Marcia Goldstein
200 Park Avenue
New York, NY   10017
   (212) 697-5083

Norfolk and Portsmouth Bar Association
c/o William Davis
1105 Virginia National Bank
Norfolk, VA   23510
   (804) 622-3152

Oneida County Bar Association
c/o Frank Nebush, Jr.
502 B Mayro Building
Utica, NY   13501
   (315) 724-3158

Onondaga County Bar Association
c/o Eloise Smarrelli
1015 State Tower Building
Syracuse, NY   13201
   (315) 471-2667

Orange County Bar Association
c/o Richard Lytle
17291 Irvine Boulevard
Tustin, CA   92680
   (714) 838-9200

Palo Alto Area Association
c/o Betty Schindler
405 Sherman Avenue
Palo Alto, CA   94306
   (415) 326-8322

Prince George County Bar Association
c/o John Joyce
6401 New Hampshire Avenue
Hyattsville, MD   20783
   (301) 270-2800

Queens County Bar Association
c/o Fred Brue
90-35 148th Street
Jamaica, NY   11435
   (212) 291-4504

Ramsey County Bar Association
c/o Jane Harens
40 North Milton Street
Suite 105

St. Paul, MN   55118
   (612) 222-0846

Richmond Bar Association
c/o Hunter Martin
1002 Mutual Building
Richmond, VA   23219
   (804) 643-8616

Richmond County Bar Association
c/o Richard Lasher
2012 Victory Boulevard
Staten Island, NY   10304
   (212) 447-5353

Riverside County Bar Association
c/o Charlotte Lewis
3765 Tenth Street
Riverside, CA   92501
   (714) 682-7520

Roanoke Bar Association
c/o James Kincanon
720 Shenandoah Building
Roanoke, VA   24011
   (703) 344-8722

Rockland County Bar Association
c/o Elleen Lutz
60 South Main Street
New City, NY   10956
   (914) 634-2149

Sacramento County Bar Association
c/o Roxanne Summers
901 H Street
Suite 101
Sacramento, CA   95814
   (916) 448-1087

San Bernardino County Bar Association
c/o Lowell Jamison
364 North Arrowhead Avenue
San Bernardino, CA   92401
   (714) 888-6791

San Diego County Bar Association
c/o Julie Hegg
1200 Third Avenue
Suite 604
San Diego, CA   92101
   (714) 231-0781

San Fernando Valley Bar Association
c/o Susan Keating
14328 Victory Boulevard
Suite 208
Van Nuys, CA 91401
(213) 786-5055

San Francisco Bar Association
c/o Irving Reichert
220 Bush Street
21st Floor
San Francisco, CA 94104
(415) 392-3960

San Francisco Lawyers Club
c/o Judee Barns
1255 Post Street
San Francisco, CA 94109
(415) 673-6025

San Joaquin County Bar Association
c/o Adelle Barrette
301 East Weber Avenue
Stockton, CA 95202
(209) 948-4620

San Mateo County Bar Association
c/o Ramon Lelli
333 Bradford Street
Redwood City, CA 94063
(415) 369-4149

Santa Clara County Bar Association
c/o Gretchen Blood
111 North Market Street
712
San Jose, CA 95113
(408) 288-8840

Seattle-King County Bar Association
c/o Helen Geisness
320 Central Building
Seattle, WA 98104
(206) 623-2551

Sierra Club Legal Defense Fund
311 California Street
Suite 311
San Francisco, CA 94104
(415) 398-1411
Executive Director: Fredric
Sutherland

Sierra Club Legal Defense Fund
Rocky Mountain Office
1612 Tremont Place
Suite 335
Denver, CO 80202
(303) 892-6301
Director: Anthony Ruckel

Sierra Club Legal Defense Fund
Southeast Alaska Office
419-6th Street
Suite 321
Juneau, AK 99801
(907) 586-2751
Staff Attorney: Stephan Volker

Sierra Club Legal Defense Fund
Washington DC Office
1424 K Street, N.W.
Suite 600
Washington, DC 20005
(202) 347-1770
Director: Khristine Hall

South Bay Bar Association
c/o Ann Gonzales
826 Maple Avenue
Torrance, CA 90503
(213) 320-4295

South Dakota State Bar
c/o William Sahr
222 East Capitol
Pierre, SD 57501
(605) 224-7554

Southeast District Bar Association
c/o Richard Lapan
8050 East Florence
Downey, CA 90241
(213) 923-7206

Spokane County Bar Association
Paulson Building
Spokane, WA 99201
(509) 747-8658
Director

Suffolk County Bar Association
c/o Sandra Lewis
4175 Veterans Memorial Highway
Ronkonkoma, NY 11779

(516) 981-1600

U.S.
Justice Department
Pollution Control Section
9th and Pennsylvania Avenue, N.W.
Washington, DC 20530
   (202) 739-2709
   Chief: Alfred Ghorzi

Ventura County Bar Association
c/o Vivian Stephenson
141 West Second Street
Oxnard, CA 93030
   (805) 486-5715

Virginia Bar Association
c/o Ward Sims
P.O. Box 5205
Charlottesville, VA 22903
   (804) 977-1396

Virginia State Bar
c/o Samuel Clifton
700 East Main Street
Richmond, VA 23219
   (804) 786-2061

Washington State Bar Association
c/o Edward Friar
505 Madison
Seattle, WA 98104
   (206) 622-6054

Westchester County Bar Association
c/o Ellen Cherry
65 Court Street
White Plains, NY 10601
   (914) 761-3707

White Plains Bar Association
c/o Leslie Levine
175 Main Street
White Plains, NY 10601
   (914) 761-2033

Wisconsin State Bar
c/o James Hough
402 West Wilson Street
Madison, WI 53703
   (608) 257-3838

Worcester County Bar Association
c/o John Moynihan
390 Main Street
Worcester, MA 01608
   (617) 791-8181

PHYSICAL DIRECTORIES

3. AIR-RELATED AGENCIES AND ORGANIZATIONS

Air Pollution Control Association
4400 Fifth Avenue
Pittsburg, PA 15213
  (412) 621-1090
  President: Dr. Richard Boubel

Alabama
Air Pollution Control Commission
645 South McDonough Street
Montgomery, AL 36104
  (205) 832-6770
  Director: James Cooper

Alaska
Department of Environmental Conser-
  vation
Air Quality
Pouch O
Juneau, AK 99811
  (907) 465-2667
  Stan Hungerford

American Lung Association
1740 Broadway
New York, NY 10019
  (212) 245-8000
  Director: Charles Kiesewetter

Arizona
Department of Health Services
Bureau of Air Pollution Control
1740 West Adams Street
Phoenix, AZ 85007
  (602) 255-1024

Arkansas
Department of Pollution Control and
  Ecology
Air Division
80001 National Drive
Little Rock, AR 72209
  (501) 371-1701
  Chief: Rick McCabe

California
Air Resources Board
1800 R Street
Sacramento, CA 95814
  (916) 445-0960
  Jerry Shiebe

Colorado
Department of Health
Air Pollution Control Division
4210 East 11th Avenue
Denver, CO 80220
  (303) 320-8333
  Director: James Lents

Connecticut
Department of Environmental Protection
Air Compliance Unit
State Office Building
165 Capitol Avenue
Hartford, CT 06115
  (203) 566-5599
  Director: Leonard Bruckman

Delaware
Department of Natural Resources
Division of Environmental Control
Air Resources Section
Tatnall Building
P.O. Box 1401
Dover, DE 19901
    (302) 678-4791
    Manager: Robert French

Delaware Valley Citizens Council
    for Clean Air
311 South Juniper Street
Philadelphia, PA   19107
    (215) 545-1832
    Kevin Quinn

District of Columbia
Department of Environmental Services
Bureau of Air and Water Quality
5010 Overlook Avenue, S.W.
Washington, DC 20032
    (202) 767-7486
    Chief: John Brink

Environmental Law Institute
Air and Water Programs
1346 Connecticut Avenue, N.W.
Suite 600
Washington, DC 20036
    (202) 452-9600
    Director: Phillip Reed

Florida
Department of Environmental Reg-
    ulations
Air Quality Management Bureau
2562 Executive Center Circle East
Montgomery Building
Tallahassee, FL 32301
    (904) 488-1344
    Chief: Dr. J.P. Subramani

Georgia
Department of Natural Resources
Environmental Protection Division
Air Protection Branch
270 Washington Street, S.W.
Room 822
Atlanta, GA 30334
    (404) 656-6900
    Chief: Robert Collom

Guam
Environmental Protection Agency
P.O. Box 2999
Agana, GU 96910
    646-7916
    Administrator: O.V. Natarajan

Hawaii
State Department of Health
Environmental Programs
P.O. Box 3378
Honolulu, HI 96801
    (808) 548-4139
    Deputy Director: Dr. James Kumagai

Idaho
Department of Health and Welfare
Boise, ID 83720
    (208) 384-2336
    Director: Murray Michael

Illinois
Environmental Protection Agency
Division of Air Pollution Control
200 West Washington Street
Springfield, IL 62701
    (217) 782-7327
    Manager: Dan Goodwin

Indiana
State Board of Health
Division of Air Pollution Control
1330 West Michigan Street
Indianapolis, IN 46206
    (317) 633-0600

Iowa
Department of Environmental Quality
Air and Land Quality Management
Wallace Building
900 East Grand
Des Moines, IA   50319

Kansas
Department of Health and Environment
Division of Environment
Bureau of Air Quality
Topeka, KS 66620
    (913) 862-9360
    Director: Howard Saiger

Kentucky
Department for Natural Resources and
   Environmental Protection
Division of Air Pollution Control
West Frankfort Office Complex
U.S. 127 South
Frankfort, KY 40601
   (502) 564-3382
   Director: Norman Schell

Louisiana
Air Control Commission
325 Loyola Avenue
New Orleans, LA 70160
   (504) 568-5100
   Technical Secretary: James Coerver

Maine
Department of Environmental Protec-
   tion
Bureau of Air Quality Control
State House
Augusta, ME 04330
   (207) 289-2437
   Chief: David Tudor

Maryland
Department of Health and Mental
   Hygiene
Air Quality Control Programs
201 West Preston Street
Baltimore, MD 21201
   (301) 383-2779
   Administrator: George Ferreri

Massachusetts
Executive Office of Environmental
   Affairs
Department of Environmental Quality
Division of Air Quality
600 Washington Street
Boston, MA 02111
   (617) 727-2658
   Director: Dr. Anthony Cortese

Metropolitan Washington Coalition
   for Clean Air
1714 Massachusetts Avenue, N.W.
Washington, DC 20036
   (202) 785-2444
   President: Richard Pardo

Michigan
Department of Natural Resources
Air Quality Division
Box 30028
Lansing, MI 48909
   (517) 373-1220
   Chief: Delbert Rector

Minnesota
Pollution Control Agency
Division of Air Quality
1935 West County Road B2
Roseville, MN 55113
   (612) 296-7331
   Director: Edward Wiik

Mississippi
Air and Water Pollution Control
   Commission
Division of Air Pollution Control
P.O. Box 827
Jackson, MS 39205
   (601) 354-2550
   Chief: Dwight Wylie

Missouri
Division of Environmental Quality
Air Quality Program
P.O. Box 1368
Jefferson City, MO 65101
   (314) 751-3241
   Director: Robert Schreiber, Jr.

Montana
Department of Health and Environmen-
   tal Sciences
Air Quality Bureau
Cogswell Building
Helena, MT 59601
   (406) 449-3454
   Chief: Michael Roach

National Commission on Air Quality
499 South Capitol Street, S.W.
Washington, DC 20003
   (202) 245-2405
   Director

Nebraska
Department of Environmental Control
Air Pollution Control Division

301 Centennial Mall South
P.O. Box 94947
Lincoln, NE 68509
   (402) 471-2341
   Chief: Gene Robinson

Nevada
Department of Conservation and
   Natural Resources
Air Quality
201 South Fall Street
Carson City, NV 89710
   (702) 885-4670
   Officer: Richard Serdoz

New Hampshire
Department of Health and Welfare
Air Pollution Control Agency
State Laboratory Building
Hazen Drive
Concord, NH 03301
   (603) 271-2281
   Director: Dennis Lunderville

New Mexico
Environmental Improvement Agency
Air Quality Division
P.O. Box 2348
Santa Fe, NM 87503
   (505) 827-5271
   Chief: Kenneth Hargis

New Jersey
Department of Environmental Pro-
   tection
Division of Environmental Quality
Bureau of Air Pollution Control
P.O. Box CN027
Trenton, NJ 08625
   (609) 292-6704
   Chief: Herbert Wortreich

New York
Department of Environmental Conservation
Division of Air Resources
50 Wolf Road
Albany, NY 12233
   (518) 457-7231

North Carolina
Department of Natural Resources

Division of Environmental Management
Air Quality Section
P.O. Box 27687
Raleigh, NC 27611
   (919) 733-4058
   Chief: James McColman

North Dakota
Department of Health
Division of Environmental Engineering
Missouri Office Building
Bismark, ND 58505
   (701) 224-2374
   Director: Gene Christianson

Ohio
Environmental Protection Agency
Office of Air Pollution Control
P.O. Box 1049
Columbus, OH 43216
   (614) 466-6686
   Acting Chief: Chuck Taylor

Oklahoma
Department of Health
Air Quality Service
P.O. Box 53551
Oklahoma City, OK 73105
   (405) 271-5220
   Chief: John Gallion

Oregon
Department of Environmental Quality
522 S.W. 5th Avenue
P.O. Box 1760
Portland, OR 97207
   (503) 229-6403
   Air Quality Coordinator: Howard
      Harris

Pennsylvania
Department of Environmental Resources
Bureau of Air Quality and Noise
P.O. Box 2063
Harrisburg, PA 17120
   (717) 787-4324
   Director: James Hambright

Puerto Rico
Environmental Quality Board
Air and Water
P.O. Box 11488

Santurce, PR 00910
  (809) 725-8692
  Vice Chairman: Pedro Marrero

Rhode Island
Department of Environmental Manage-
  ment
Division of Air Pollution Control
Health Building
Davis Street
Room 204
Providence, RI 02908
  (401) 277-2808
  Chief: Thomas Wright

South Carolina
Department of Health and Environ-
  mental Control
Bureau of Air Quality Control
2600 Bull Street
Columbia, SC 29201
  (803) 758-5406
  Chief: William Crosby

South Dakota
Department of Environmental Pro-
  tection
Air Quality and Solid Waste
Foss Building
Pierre, SD 57501
  (605) 224-3329
  Chief: Joel Smith

Tennessee
Department of Public Health
Bureau of Environmental Health
Division of Air Pollution Control
256 Capitol Hill Building
Nashville, TN 37219
  (615) 741-3931
  Director: Harold Hodges

Texas
Air Control Board
8520 Shoal Creek Boulevard
Austin, TX 78758
  (512) 451-5711
  Executive Director: Bill Stewart

U.S.
Department of Commerce

National Oceanic and Atmospheric
  Administration
Washington, DC 20230
  (202) 443-8910
  Administrator: Richard Frank

U.S.
Department of Health, Education and
  Welfare
National Center for Air Pollution Con-
  trol Advisory Committee
330 Independence Avenue, S.W.
Washington, DC 20201
  (202) 377-2000
  Director

U.S.
Environmental Protection Agency
Air, Noise and Radiation
401 M Street, S.W.
Washington, DC 20460
  (202) 755-2673
  David Hawkins

U.S.
Environmental Protection Agency
Air Quality Planning and Standards
401 M Street, S.W.
Washington, DC 20460
  (202) 755-2673
  Deputy Assistant Administrator:
    Walter Barber

Utah
State Division of Health
Bureau of Air Quality
44 Medical Drive
Salt Lake City, UT 84113
  (801) 533-6111
  Director: Alvin Rickers

Vermont
Agency of Environmental Conservation
Air Pollution
Montpelier, VT 05602
  (802) 828-3395
  Richard Valentinetti

Virginia
State Air Pollution Control Board
Operations and Procedures

1106 Ninth Street Office Building
Richmond, VA 23219
  (804) 786-7564
  Director: James Ruehrmund

Virgin Islands
Department of Conservation and
  Cultural Affairs
Division of Natural Resources
P.O. Box 578
St. Thomas, VI 00801
  (809) 774-6420
  Associate Director: Don Francois

Washington
Department of Ecology
Air Monitoring Branch
Olympia, WA 98504
  (206) 753-2800
  Stu Clark

Washington
Department of Ecology
Air Resources Management Division
Olympia, WA 98504
  (206) 753-2800
  Division Supervisor: Henry Droese

West Virginia
Air Pollution Control Commission
1558 East Washington Street
Charleston, WV 25311
  (304) 348-3286
  Director: Carl Beard, II

Wisconsin
Department of Natural Resources
Division of Environmental Standards
Bureau of Air Management
P.O. Box 7921
Madison, WI 53707
  (608) 266-0603
  Director: Dr. Robert Arnott

Wyoming
Department of Environmental Quality
Air Quality Division
Hathaway Building
Cheyenne, WY 82002
  (307) 777-7391
  Administrator: Randolph Wood

4. EARTH-RELATED AGENCIES AND ORGANIZATIONS

Alaska Association of Soil Conser-
vation Subdistricts
Box 274
Palmer, AK 99645
  (907) 745-3681
  President: Wayne Bouwens

American Federation of Mineralogical
and Lapidary Societies
4727 North 24th Street
Arlington, VA 22207
  (703) 528-8093
  President: Kenneth Zahn

American Geological Institute
5205 Leesburg Pike
Falls Church, VA 22041
  (703) 379-2480
  John Mulvhill

Arizona
Bureau of Geology and Mineral Tech-
nology
University of Arizona
Tucson, AZ 85721
  (602) 626-1943
  Director: Dr. W.H. Dresher

Colorado Association of Soil Conservation Districts
4644 Weld Co. Road 20
Longmont, CO 80501
  (303) 776-5516
  President: Leo Berger

Delaware
Geological Survey
University of Delaware
Newark, DE 19711
  (302) 738-2833
  State Geologist: Robert Jordan

Georgia
Department of Natural Resources
Geological Survey
270 Washington Street, S.W.
Atlanta, GA 30334
  (404) 656-3530
  Chief: Joseph Murray

Idaho Association of Soil Conservation
Districts
Route 2
Box 2190
Nampa, ID 83651
  (208) 466-5819
  President: Lowell Grim

Indiana
Department of Natural Resources
State Geological Survey
611 North Walnut Grove
Bloomington, IN 47401
  (812) 337-2862
  Director: John Patton

Iowa Association of Soil Conservation
District Commissioners
Route 2

Britt, IA 50423
  (515) 565-3339
  President: Charles McLaughlin

Maryland Association of Soil Conser-
  vation Districts
Parole Plaza Office Building
Annapolis, MD 21401
  (301) 269-2338
  President: William Sutton

Maryland
Department of Natural Resources
Geological Survey
John Hopkins University
Baltimore, MD 21218
  (301) 235-0771
  Director: Dr. Kenneth Weaver

Maryland
State Soil Conservation Committee
Parole Plaza Office Building
Annapolis, MD 21401
  (301) 269-2338
  Chairman: Vernon Foster

Michigan
Department of Natural Resources
Geological Survey Division
Box 30028
Lansing, MI 48909
  (517) 373-1220
  Chief: Arthur Slaughter

Michigan Soil Conservation Districts
Route 1
Linwood, MI 48634
  (517) 879-2755
  President: Elmer Lambert

Minnesota
Geological Survey
University of Minnesota
1633 Eustis Street
St. Paul, MN 55108
  (612) 373-3372
  Director: Matt Walton

Montana
Bureau of Mines and Geology
c/o Montana College of Mineral
  Science and Technology

Butte, MT 59701
  (406) 792-8321
  Director: S.L. Groff

Nebraska
Geological Survey
Conservation and Survey Division
University of Nebraska
Lincoln, NE 68508
  (402) 472-3471
  Director: V.H. Dreeszen

Nevada
Bureau of Mines and Geology
University of Nevada
Reno, NV 89557
  (702) 784-6691
  State Geologist: John Schilling

New Jersey
Bureau of Geology and Topography
Box 1390
Trenton, NJ 08625
  (609) 292-2576
  State Geologist: Kemble Widmer

New Jersey
Department of Agriculture
State Soil Conservation Committee
P.O. Box 1888
Trenton, NJ 08625
  (609) 292-5540
  Executive Secretary: Samuel Race

New York
Geological Survey
State Museum and Science Service
State Education Building
Room 973
Albany, NY 12234
  (518) 474-5816
  State Geologist: Vacant

New York Soil Conservation Districts
  Association
North Creek, NY 12852
  (518) 251-2619
  President: William Waddell

North Carolina
Department of Natural Resources and
  Community Development

Earth Resources
P.O. Box 27687
Raleigh, NC 27611
   (919) 733-4984
   Stephen Conrad

North Dakota Association of Soil
  Conservation Districts
New Rockford, ND 58356
   (701) 947-5490
   President: William Starke

Seismological Society of America
2620 Telegraph Avenue
Berkeley, CA 94704
   (415) 848-0954
   Director: Susan Newman

Soil Conservation Society of America
7515 N.E. Ankeny Road
Ankeny, IA 50021
   (515) 289-2331
   President: Arthur Latornell

South Dakota
Department of Natural Resource
  Development
Division of Geological Surveying
Room 301
Akeley Science Center
University of South Dakota
Vermillion, SD 57069
   (605) 624-4471
   Director: Duncan McGregor

Texas
Bureau of Economic Geology
University of Texas at Austin
University Station
Box X
Austin, TX 78712
   (512) 471-1534
   Director: W.L. Fisher

U.S.
Department of Agriculture
Science and Education Department
Watershed Hydrology
Washington, DC 20250
   (301) 344-4240
   Director: Dr. David Farrell

U.S.
Department of Agriculture
Soil Conservation Service
P.O. Box 2890
Washington, DC 20013
   (202) 447-4531
   Administrator: R.M. Davis

U.S.
Department of Agriculture
Soil Conservation Service
Ecological Sciences Staff
P.O. Box 2890
Washington, DC 20013
   (202) 447-4531

U.S.
Department of the Interior
Geological Survey
Central Regional Office
Box 25046
Denver Federal Center
Denver, CO 80225
   (303) 234-4630
   Director

U.S.
Department of the Interior
Geological Survey
Eastern Regional Office
109 National Center
Reston, VA 22092
   (703) 860-7414
   Director

U.S.
Department of the Interior
Geological Survey
EIA Program
Reston, VA 22092
   (703) 860-7414
   Director

U.S.
Department of the Interior
Geological Survey
Geologic Division
910 National Center
12201 Sunrise Valley Drive
Reston, VA 22092
   (703) 860-6531
   Chief Geologist

U.S.
Department of the Interior
Geological Survey
Information Office
119 National Center
Reston, VA   22092
   (703) 860-7444
   Information Officer

U.S.
Department of the Interior
Geological Survey
National Center
12201 Sunrise Valley Drive
Reston, VA   22092
   (703) 860-7444
   Director: H.W. Menard

U.S.
Department of the Interior
Geological Survey
U.S. National Earthquake Information
   Service
Box 25046
Denver Federal Center
Denver, CO   80225
   (303) 234-3994
   NEIS Coordinator

U.S.
Department of the Interior
Geological Survey
Western Regional Office
345 Middlefield Road
Menlo Park, CA   94025
   (415) 323-2711
   Director

U.S.
Nuclear Regulatory Commission
Division of Site Safety and Environ-
   mental Analysis
Geoscience Branch
Washington, DC   20555
   (301) 492-7972
   Chief: J.Carl Stepp

Utah Association of Soil Conservation
   Districts
8915 South 700
East Sandy, UT   84070

   (801) 255-3642
   President: Jim Jatsumori

Utah
Environmental Coordinating Committee
Division of Geological and Mineral
   Survey
124 State Capitol
Salt Lake City, UT   84111
   (801) 533-5794
   Bruce Kaliser

Utah
State Department of Natural Resources
Geological and Mineral Survey
606 Black Hawk Way
Salt Lake City, UT   84108
   (801) 581-6831
   Director: Donald McMillan

Utah State Soil Conservation
   Commission
147 North 200 West
Salt Lake City, UT   84103
   (801) 533-5421
   Executive Secretary: James Harvey

Vermont
Agency of Environmental Conservation
State Geologist
Montpelier, VT   05602
   (802) 828-3357
   State Geologist: Charles Ratte

Washington
Department of Natural Resources
Geology and Earth Resources
Public Lands Building
Olympia, WA   98504
   (206) 753-5327
   Division Supervisor: Vaughn Living-
   ston, Jr.

Wisconsin
Geological and Natural History Survey
University of Wisconsin
1815 University Avenue
Madison, WI   53706
   (608) 262-1705
   Director: Meredith Ostrom

5. NOISE-RELATED AGENCIES AND ORGANIZATIONS

Acoustical Society of America
335 East 45th Street
New York, NY 10017
   (212) 661-9404
   Administrative Secretary

Arkansas
Department of Pollution Control and
   Ecology
8001 National Drive
Little Rock, AR 72209
   (501) 371-1701
   Assistant Director: Bobby Voss

California
State Department of Health
Office of Noise Control
2151 Berkeley Way
Berkeley, CA 94704
   (415) 843-7900
   Chief: A.E. Lowe

Colorado
Department of Health
4210 East 11th Avenue
Denver, CO 80220
   (303) 388-6111
   Noise Control Officer: Belmont
   Evans

Connecticut
Department of Environmental Pro-
   tection
Office of Noise Control
State Office Building

165 Capitol Avenue
Hartford, CT 06115
   (203) 566-4855
   Director: Vacant

Delaware
Department of Natural Resources and
   Environmental Control
Division of Environmental Control
Air Resources Section
P.O. Box 1401
Dover, DE 19901
   (302) 678-4791
   Manager: Robert French

District of Columbia
Department of Environmental Services
Environmental Health Administration
415 12th Street, N.W.
Room 314
Washington, DC 20004
   (202) 724-4358
   Marvin Fink

Florida
Department of Environmental Regula-
   tions
2600 Blair Stone Road
Tallahassee, FL 32301
   (904) 487-2095
   Noise Section Administrator: Jesse
   Borthwick

Georgia
Department of Human Resources

Division of Physical Health
Special Operations Unit
47 Trinity Avenue, S.W.
Room 313-H
Atlanta, GA 30334
   (404) 656-4660
   Chief: Charles Head

Indiana
State Board of Health
Industrial Hygiene
1330 West Michigan Street
Indianapolis, IN 46206
   (317) 633-0146
   Director: Virgil Konopinski

Kentucky
Department for Natural Resources and
   Environmental Protection
Division of Special Programs
Century Plaza
U.S. 127 South
Frankfort, KY 40601
   (502) 564-7274
   Noise Control Officer: Tommie
      Jackson

Louisiana
Department of Health and Human
   Resources
Occupational Health
P.O. Box 60630
New Orleans, LA 70160
   (504) 568-5139
   Administrator: Wilfred Charbonnet

Maine
Department of Human Services
Division of Health Engineering
State House
Augusta, ME 04330
   (207) 289-3826
   Director: Donald Hoxie

Maryland
Division of Noise Control
201 West Preston Street
Baltimore, MD 21201
   (301) 383-2727
   Chief: Thomas Towers

Massachusetts
Department of Public Health
Division of Environmental Health
Air Pollution Control
600 Washington Street
Boston, MA 02111
   (617) 727-2658
   Associate: Elise Comproni

Michigan
Department of Natural Resources
Office of Environmental Enforcement
Mason Building
Lansing, MI 48909
   (517) 373-3503
   Chief: Jack Bailf

Minnesota
Pollution Control Agency
Division of Air Quality and Technical
   Services
1935 West County Road B2
Roseville, MN 55113
   (612) 296-7340
   Al Perez

Mississippi
2003 Sillers Building
500 High Street
Jackson, MS 39201
   (601) 354-7575
   Federal-State Coordinator: James
      Fleming

Montana
Department of Health and Environmental
   Sciences
Environmental Sciences Division
Occupational Health Program
Cogswell Building
Helena, MT 59601
   (406) 449-3946
   Chief: Larry Lloyd

Nevada
Department of Conservation and Natural
   Resources
201 South Fall Street
Carson City, NV 89710
   (702) 885-4670
   Environmental Specialist: Hugh Ricci

New Hampshire
Department of Health and Welfare
Occupational Health Services
State Laboratory Building
Hazen Drive
Concord, NH 03301
  (603) 271-2281
  Max Helgemeier

New Jersey
Department of Environmental Protection
Office of Noise Control
P.O. Box 1390
Trenton, NJ 08625
  (609) 292-7696
  Director: Edward DiPolvere

New Mexico
Health and Environment Department
Noise Control Unit
Crown Building
P.O. Box 968
Sante Fe, NM
  (505) 827-5271
  Coordinator: James Libberton

New York
Department of Environmental Con-
  servation
Bureau of Noise Control
50 Wolf Road
Albany, NY 12233
  (518) 457-6603
  Director: Dr. Fred Haag

North Carolina
Department of Natural Resources and
  Community Development
P.O. Box 27687
Raleigh, NC 27603
  (919) 733-4984
  Secretary: Howard Lee

Ohio
Department of Public Health
246 North High Street
Columbus, OH 43216
  (614) 466-1390
  George McClain

Oklahoma
State Department of Health

Occupational RAD Health Services
N.E. 10th and Stonewall
Oklahoma City, OK 73105
  (405) 271-5221
  Chief: Dale McHard

Oregon
Department of Environmental Quality
Noise Control
522 S.W. 5th Avenue
P.O. Box 1760
Portland, OR 97207
  (503) 229-6403
  Coordinator: John Hector

Puerto Rico
Environmental Quality Board
Noise and Solid Waste
P.O. Box 11488
Santurce, PR 00910
  (809) 725-5140
  Associate Member: Santos Rohena

Rhode Island
Department of Health
Division of Occupational Health
206 Health Building
Davis Street
Providence, RI 02903
  (401) 277-2438
  Chief: Dr. James Deery

South Carolina
Department of Health and Environmental
  Control
Division of Occupational Health
2600 Bull Street
Columbia, SC 29201
  (803) 758-5681
  Director: Dr. H.G. Callison, Jr.

Texas
State Department of Health
Division of Occupational Health and
  Radiation Control
1100 West 49th Street
Austin, TX 78756
  (512) 458-7341
  Director: David Lacker

U.S.
Environmental Protection Agency

Noise Abatement and Control
401 M Street, S.W.
Washington, DC 20460
  (202) 755-2673
  Deputy Assistant Administrator:
    Charles Elkins

<u>Utah</u>
State Division of Health
Radiation and Occupational Health
  Section
150 WM Temple
Salt Lake City, UT 84113
  (801) 533-6734
  Chief: Larry Anderson

<u>Vermont</u>
Agency for Environmental Conservation
Environmental Engineering
State Office Building
Montpelier, VT 05602
  (802) 826-3361
  Director: Reginald LaRosa

<u>Virginia</u>
Department of Health
Bureau of Occupational Health
109 Govenor Street
Room 923
Madison Building
Richmond, VA 23219
  (804) 786-6285
  Director: Charles Harrigan

<u>West Virginia</u>
Department of Health
Bureau of Industrial Hygiene
151 11th Avenue
South Charleston, WV 25303
  (304) 348-3526
  Director: William Aaroe

6. PLANT/ANIMAL-RELATED AGENCIES AND ORGANIZATIONS

Academy of Natural Sciences
Systematic Biology Division
19th Street and the Parkway
Philadelphia, PA 19103
  (215) 299-1180
  Director

Alabama
Department of Agriculture and
  Industries
Division of Agricultural Chemistry
P.O. Box 33356
Montgomery, AL 36109
  (205) 832-3700
  Director: Dr. John Bloch

Alabama Ornithological Society
134 Woodland Hills
Tuscaloosa, AL 35401
  President: Dr. James Thompson

Alabama Wildlife Federation
660 Adams Avenue
Montgomery, AL 36104
  (205) 263-6565
  President: Walter Ernest

Alaska
Department of Environmental Conser-
  vation
P.O. Box 1088
Palmer, AK 99645
  (907) 745-4686
  Pesticides Specialist: Dr. William
    Burgoyne

Alaska Wildlife Federation
Box 3072
Route 3
Juneau, AK 99801
  (907) 586-6114
  President: Ed Gustafson

American Fisheries Society
5410 Grosvenor Lane
Bethesda, MD 20014
  (301) 897-8616
  President: Dr. Henry Regier

American Forest Institute
1619 Massachusetts Avenue, N.W.
Washington, DC 20036
  (202) 797-4500
  President: John Ball

American Forestry Association
1319 18th Street, N.W.
Washington, DC 20036
  (202) 467-5810
  President: Carl Reidel

American Institute of Fishery Research
  Biologists
1226 Skyline Drive
Edmonds, WA 98020
  (206) 774-1798
  President: Eugene Nakamura

American Institute of Landscape
  Architects
6810 North Second Place

Phoenix, AZ 85012
  (602) 277-0096
  Executive Vice President: F.J.
    MacDonald

American Ornithologists' Union
National Museum of Natural History
Smithsonian Institution
Washington, DC 20560
  (212) 873-1300
  President: Wesley Lanyon

American Society of Landscape
  Architects
1900 M Street, N.W.
Washington, DC 20036
  (202) 466-7730
  President: Lane Marshall

American Society of Mammalogists
Museum of Natural History
University of Kansas
Lawrence, KS 66045
  (913) 864-3673
  President: Robert Hoffmann

American Society of Zoologists
Box 2739
California Lutheran College
Thousand Oaks, CA 91360
  (805) 492-4055
  President: Mary Rice

Anchorage Audubon Society
P.O. Box 1161
Anchorage, AK 99510
  (907) 694-3503
  President: Lou Carufel

Arizona
Cooperative Fishery Research Unit
210 Biological Sciences East
University of Arizona
Tucson, AZ 85721
  (602) 626-1959
  Leader: Dr. Jerry Tash

Arizona
Cooperative Wildlife Research Unit
University of Arizona
Tucson, AZ 85721

  (602) 626-1193
  Leader: Dr. Lyle Sowls

Arizona Council for Environmental
  Studies
University of Arizona
Tucson, AZ 85721
  (602) 884-3197
  Director: Dr. David Byrne

Arizona
Game and Fish Department
2222 West Greenway Road
Phoenix, AZ 85023
  (602) 942-3000
  Director: Robert Jantzen

Arizona Wildlife Federation
Box 27573
Phoenix, AZ 85061
  (602) 264-3884
  Executive Director: Thomas Sullivan

Arkansas Audubon Society
2600 Riviera Circle
Fort Smith, AR 72903
  (501) 452-3059
  President: Carol Wooten

Arkansas
State Plant Board
Division of Feeds, Fertilizer and
  Pesticides
P.O. Box 1069
Little Rock, AR 72203
  (501) 371-1021
  Director: Ralph Pay

Arkansas Wildlife Federation
P.O. Box 2661
Little Rock, AR 72203
  (501) 375-6161
  President: Nesbit Bowers

Association of Consulting Foresters
Box 6
Wake, VA 23176
  (804) 776-4031
  President: Edward Bailey

Association of Midwest Fish and
  Wildlife Agencies
Fish and Wildlife Division
Department of Natural Resources
607 State Office Building
Indianapolis, IN 46204
  (317) 633-7696
  President: Frank Lockard

Audubon Council of Illinois
320 South Third Street
Rockford, IL 61108
  (815) 964-6666
  President: Gerald Paulson

Audubon Society of Missouri
1308 Wilson Avenue
Columbia, MO 65201
  (314) 443-8946
  President: Jim Rathert

Audubon Society of New Hampshire
3 Silk Farm Road
Concord, NH 03301
  (603) 224-9909
  President: Jane Grant

Audubon Society of Rhode Island
40 Bowen Street
Providence, RI 02903
  (401) 521-1670
  President: Mason Cocroft

California
Department of Fish and Game
1416 Ninth Street
Sacramento, CA 95814
  (916) 445-1383
  John Turner

California
Department of Food and Agriculture
1220 N Street
Sacramento, CA 95814
  (916) 322-1992
  Gordon Snow

California
Department of Forestry
1416 Ninth Street
Room 1506-17

Sacramento, CA 95814
  (916) 445-9144
  Cliff Chapman

California Forest Protective
  Association
1127 11th Street
Room 534
Sacramento, CA 95814
  (916) 444-6592
  Executive Vice President: Stanley
  Hulett

California Native Plant Society
2380 Ellsworth Street
Suite D
Berkeley, CA 94704
  (415) 841-5575
  President: James Smith, Jr.

California Natural Resources Federation
2775 Cottage Way
Suite 39
Sacramento, CA 95825
  (916) 483-1125
  President: Rudolph Schafer

California Wildlife Federation
P.O. Box 9504
Sacramento, CA 95823
  (916) 447-4025
  President: Loren Lutz

Cary Arboretum of the New York
  Botanical Gardens
Box AB
Millbrook, NY 12545
  (914) 677-5343
  Director: Dr. Willard Payne

Center for Natural Areas
Forest Ecology
1525 New Hampshire Avenue, N.W.
Washington, DC 20036
  (202) 265-0066
  Forest Ecologist: David Tilles

Center for Natural Areas
Wildlife Biology
1525 New Hampshire Avenue, N.W.

Washington, DC 20036
   (202) 265-0066
   Wildlife Biologist: James Hynson

Colorado
Department of Agriculture
Division of Plant Industry
State Services Building
1525 Sherman Street
Denver, CO 80203
   (303) 892-2838
   Director: Robert Sullivan

Colorado Wildlife Federation
5725 St. Vrain Road
Longmont, CO 80501
   (303) 776-3999
   President: David Merrifield

Connecticut Audubon Council
20 Union Street
Seymour, CT 06483
   (203) 888-3124
   President: Stephen Davis

Connecticut Audubon Society
2325 Burr Street
Fairfield, CT 06430
   (203) 259-6305
   President: Robert Larsen

Connecticut
Department of Agriculture
Room 273
State Office Building
Hartford, CT 06115
   (203) 566-4667
   Commissioner: Leonard Krogh

Connecticut
Department of Environmental Pro-
   tection
Fish and Wildlife Unit
State Office Building
165 Capitol Avenue
Hartford, CT 06115
   (203) 566-5599
   Chief: Cole Wilde

Connecticut
Department of Environmental Protection

Hazardous Waste Management Unit
State Office Building
165 Capitol Avenue
Hartford, CT 06115
   (203) 566-5148
   Director: Dr. Stephen Hitchcock

Connecticut Wildlife Federation
P.O. Box 7
Middletown, CT 06457
   (203) 347-1291
   President: John Reilly, III

Defenders of Wildlife
1244 19th Street, N.W.
Washington, DC 20036
   (202) 659-9510
   President: Jocelyn Alexander

Delaware
Department of Agriculture
Drawer D
Dover, DE 19901
   (302) 678-4811
   Secretary: W.E. McDaniel

Delaware
Department of Agriculture
Forestry
Drawer D
Dover, DE 19901
   (302) 678-4820
   State Forester: Walter Gabel

Delaware
Department of Natural Resources and
   Environmental Control
Division of Fish and Wildlife
Fisheries
Tatnall Building
P.O. Box 1401
Dover, DE 19901
   (302) 678-4432
   Manager: Charles Lesser

Delaware
Department of Natural Resources and
   Environmental Control
Division of Fish and Wildlife
Wildlife
Tatnall Building
P.O. Box 1401

Dover, DE 19901
  (302) 678-4431
  Manager: Anthony Florio

Delaware Wildlife Federation
2 Connell Circle
Newark, DE 19711
  (302) 366-8518
  President: John Iorizzo

Delmarva Ornithological Society
P.O. Box 4247
Greenville, DE 19807
  President: Joanne Patterson

Desert Fishes Council
407 West Line Street
Bishop, CA 93514
  (714) 873-4095
  Chairman: Peter Sanchez

Ducks Unlimited
P.O. Box 66300
Chicago, IL 60666
  (312) 299-3334
  President: Preston Williams

Federation of New York State Bird
  Clubs
533 Chestnut Street
West Hempstead, NY 11552
  President: Stephen Dempsey

Florida Audubon Society
P.O. Drawer 7
Maitland, FL 32751
  (305) 647-2615
  Chairman: Dade Thornton

Florida
Department of Agriculture
Division of Inspection
Mayo Building
Tallahassee, FL 32304
  (904) 488-3731
  Director: Vincent Giglio

Florida Forestry Association
P.O. Box 1696
Tallahassee, FL 32302
  (904) 222-5646
  President: L.A. Woodward, Jr.

Florida Wildlife Federation
4080 North Haverhill Road
West Palm Beach, FL 33407
  (305) 683-2328
  President: Walter Brandon

Forest Farmers Association Cooperative
Suite 380
4 Executive Park East, N.E.
P.O. Box 95385
Atlanta, GA 30329
  (404) 325-2954
  President: Ralph Law

Forest History Society
109 Coral Street
Santa Cruz, CA 95060
  (408) 426-3770
  President: Harold Pinkett

Friends of Animals
11 West 60th Street
New York, NY 10023
  (212) 247-8077
  President: Alic Herrington

Fund for Animals
140 West 57th Street
New York, NY 10019
  (212) 246-2096
  President: Cleveland Amory

Georgia
Cooperative Fishery Research Unit
School of Forestry
University of Georgia
Athens, GA 30602
  (404) 546-2234
  Leader: Robert Reinert

Georgia
Department of Agriculture
Agriculture Building
Capitol Square
Atlanta, GA 30334
  (404) 656-3600
  Commissioner: Thomas Irvin

Georgia
Department of Agriculture
Pesticide Division
Agriculture Building

Capitol Square
Atlanta, GA   30334
  (404) 656-4958
  Director: J.R. Conley

Georgia
Department of Natural Resources
Fisheries Management
270 Washington Street, S.W.
Atlanta, GA   30334
  (404) 656-3530
  Chief: Mike Gennings

Georgia
Department of Natural Resources
Game Management
270 Washington Street, S.W.
Atlanta, GA   30334
  (404) 656-3530
  Chief: Terry Kile

Georgia
Forestry Commission
Box 819
Macon, GA   31202
  (912) 744-3211
  Director: Ray Shirley

Georgia Wildlife Federation
4019 Woburn Drive
Tucker, GA   30084
  (404) 934-1955
  President: Earl Wilkes

Hawaii Audubon Society
P.O. Box 22832
Honolulu, HI   96822
  (808) 262-4046
  President: Dr. Robert Pyle

Hawaii
Cooperative Fishery Research Unit
2538 The Mall
University of Hawaii
Honolulu, HI   96822
  (808) 948-8350
  Leader: Dr. James Parrish

Hawaii
Department of Agriculture
P.O. Box 22159

Honolulu, HI   96822
  (808) 548-7100
  Chairman: John Farias, Jr.

Hawaii
Department of Agriculture
Division of Plant Industry
Honolulu, HI   96822
  (808) 941-3071
  Head: Charles Yasuda

Hawaii
Department of Land and Natural
  Resources
Division of Fish and Game
1151 Punchbowl Street
Honolulu, HI   96813
  (808) 548-4000
  Director: Kenji Ego

Hawaii
Department of Land and Natural
  Resources
Division of Fish and Game
Fisheries Branch
1151 Punchbowl Street
Honolulu, HI   96813
  (808) 548-4000
  Chief: Henry Sokuda

Hawaii
Department of Land and Natural
  Resources
Division of Fish and Game
Wildlife Branch
1151 Punchbowl Street
Honolulu, HI   96813
  (808) 548-4000
  Chief: Ronald Walker

Hawaii
Department of Land and Natural
  Resources
Division of Forestry
1151 Punchbowl Street
Honolulu, HI   96813
  (808) 548-4000
  State Forester: Libezt Londgrof

Idaho
Department of Agriculture

4696 Overland Road
Boise, ID 83701
  (208) 384-3240
  Pesticide Specialist: Rod Awe

Idaho Wildlife Federation
486 Rose Street, North
Twin Falls, ID 83301
  (208) 733-4760
  President: Donald Zuck

Illinois Audubon Society
P.O. Box 441
Wayne, IL 60184
  (312) 584-6290
  President: Paul Mooring

Illinois
Department of Agriculture
Fairgrounds Emerson Building
Springfield, IL 62706
  (217) 782-2274
  Director: John Block

Illinois Wildlife Federation
Box 116
Blue Island, IL 60406
  (312) 388-3995
  President: Richard Kehn

Indiana Audubon Society
Mary Gray Bird Sanctuary
R.R. 6
Connersville, IN 47331
  (317) 825-9788
  President: Charles Moulin

Indiana
Department of Natural Resources
Division of Fish and Wildlife
608 State Office Building
Indianapolis, IN 46204
  (317) 633-6344
  Head: Frank Lockard

Indiana
Department of Natural Resources
Division of Forestry
608 State Office Building
Indianapolis, IN 46204
  (317) 633-6344
  State Forester: John Datena

Industrial Forestry Association
1220 S.W. Columbia Street
Portland, OR 97201
  (503) 222-9505
  President: W. Lee Robinson

International Association of Fish and
  Wildlife Agencies
1412 16th Street, N.W.
Washington, DC 20036
  (202) 232-1652
  President: Russell Stuart

International Society for the Protection of Animals
Field Services Office
29 Perkins Street
Boston, MA 02130
  (617) 522-7000
  Director: John Walsh

Iowa
Department of Agriculture
Pesticide Division
East 7th and Court
Des Moines, IA 50309
  (515) 281-8590
  Director: M.R. VanCleave

Iowa Ornithologists' Union
235 McClellan Boulevard
Davenport, IA 52803
  (319) 355-7051
  President: George Crossley

Iowa Wildlife Federation
P.O. Box 151
Boone, IA 50036
  (515) 432-4904
  President: Chuck Mills

Kansas Ornithological Society
R.D. 2
Box 209
Newton, KS 67114
  President: Jean Schulenberg

Kansas
State Board of Agriculture
Weed and Pesticide Division
State Office Building
Topeka, KS 66612

(913) 296-2263
Director: Freeman Biery

Kansas Wildlife Federation
R.D. 1
Wamego, KS 66547
  (913) 456-2500
  President: Walter Snell

Kentucky Audubon Council
8607 Whipps Bend Road
Louisville, KY 40222
  (502) 426-1853
  President: Ralph Madison

Kentucky
Department for Natural Resources and
  Environmental Protection
Division of Hazardous Materials
Capitol Tower Plaza
5th Floor
Frankfort, KY 40601
  (502) 564-6716
  John Smither

League of Kentucky Sportsmen
2433 Liberty Road
Lexington, KY 40505
  (606) 252-2834
  President: Joseph Coomes

League of Ohio Sportsmen
4330 Clime Road North
Columbus, OH 43228
  (614) 385-3233
  President: Harry Armstrong

Louisiana
Department of Agriculture
Office of Agriculture and Environ-
  mental Sciences
P.O. Box 16390-A
University Station
Baton Rouge, LA 70893
  (504) 342-5809
  Director: E.A. Cancienne

Louisiana Forestry Association
P.O. Drawer 5067
Alexandria, LA 71301
  (318) 443-2558
  President: Jeff Hughes, Jr.

Louisiana Wildlife Federation
Box 16089
LSU
Baton Rouge, LA 70893
  (504) 355-1871
  President: Henry Bernard, Jr.

Maine Audubon Society
Gilsland Farm
118 Route 1
Falmouth, ME 04105
  (207) 781-2330
  President

Maine
Board of Pesticide Control
Vickery Hill Building
Augusta, ME 04330
  (207) 289-2215
  Supervisor: Donald Mairs

Maryland
Department of Agriculture
Parole Plaza Building
Annapolis, MD 21401
  (301) 267-1161
  Secretary: Young Hance

Maryland
Department of Natural Resources
Fisheries Administration
Tawes State Office Building
Annapolis, MD 21401
  (301) 269-3558
  Administrator: Robert Rubelmann

Maryland
Department of Natural Resources
Forest Service
Tawes State Office Building
Annapolis, MD 21401
  (301) 269-3776
  Director: Donald MacLauchlan

Maryland Ornithological Society
Cylburn Mansion
4915 Greenspring Avenue
Baltimore, MD 21209
  (301) 224-2061
  President: James Cheevers

Maryland Wildlife Federation
415 St. Paul Street
Baltimore, MD 21202
  (301) 752-5614
  President: Paul Breidenbaugh

Massachusetts Audubon Society
South Great Road
Lincoln, MA 01773
  (617) 259-9500
  President: Francis Moulton, Jr.

Massachusetts
Cooperative Fishery Research Unit
204 Holdsworth Hall
University of Massachusetts
Amherst, MA 01003
  (413) 545-2011
  Leader: Roger Reed

Massachusetts
Cooperative Wildlife Research Unit
University of Massachusetts
Amherst, MA 01003
  (413) 545-2757
  Leader: Dr. Wendell Dodge

Massachusetts
Executive Office of Environmental
  Affairs
Department of Fisheries, Wildlife
  and Recreation Vehicles
100 Cambridge Street
Boston, MA 02202
  (617) 727-1614
  Commissioner: Bruce Gullion

Massachusetts
Executive Office of Environmental
  Affairs
Department of Fisheries, Wildlife
  and Recreation Vehicles
Division of Fisheries and Wildlife
100 Cambridge Street
Boston, MA 02202
  (617) 727-1614
  Director: Matthew Connolly

Massachusetts
Executive Office of Environmental
  Affairs

Department of Fisheries, Wildlife and
  Recreation Vehicles
Division of Marine Fisheries
100 Cambridge Street
Boston, MA 02202
  (617) 727-1614
  Director: Allen Peterson

Massachusetts
Public Health Department
Department of Environmental Quality
  Engineering
600 Washington Street
Boston, MA 02111
  (617) 727-2863
  Supervisor: Lewis Wells

Massachusetts Wildlife Federation
P.O. Q
Burlington, MA 01803
  (617) 272-0455
  President: Dr. Kalil Boghdan

Michigan
Department of Agriculture
5th Floor
Lewis Cass Building
P.O. Box 30017
Lansing, MI 30017
  (517) 373-1050
  Director: Dale Ball

Michigan
Department of Agriculture
Laboratory Division
1615 South Harrison Road
East Lansing, MI 48823
  (517) 373-1050
  Director: Dean Pridgeom

Michigan
Department of Natural Resources
Fisheries Division
Box 30028
Lansing, MI 48909
  (517) 373-1220
  Chief: John Scott

Michigan
Department of Natural Resources
Forest Management Division

Box 30028
Lansing, MI 48909
  (517) 373-1220
  Chief: Henry Webster

Michigan
Department of Natural Resources
Wildlife Division
Box 30028
Lansing, MI 48909
  (517) 373-1220

Minnesota
Department of Agriculture
420 State Office Building
St. Paul, MN 55155
  (612) 296-2856
  Commissioner: Bill Walker

Minnesota
Department of Natural Resources
Division of Fish and Wildlife
300 Centennial Building
658 Cedar Street
St. Paul, MN 55155
  (612) 296-2894
  Director: David Vesall

Minnesota
Department of Natural Resources
Division of Forestry
300 Centennial Building
658 Cedar Street
St. Paul, MN 55155
  (612) 296-0783
  Director: Rodney Sando

Minnesota
Department of Natural Resources
Ecological Section
300 Centennial Building
658 Cedar Street
St. Paul, MN 55155
  (612) 296-0783
  Supervisor: Oliver Jarvenpa

Minnesota Ornithologists' Union
James Ford Bell Museum of Natural
  History
University of Minnesota
Minneapolis, MN 55455

  (507) 532-3934
  President: Henry Kyllingstad

Mississippi
Department of Agriculture and Commerce
P.O. Box 1609
Jackson, MS 39205
  (601) 354-6563
  Commissioner: Jim Ross

Mississippi
Department of Agriculture and Commerce
Division of Plant Industry
P.O. Box 5207
Mississippi State, MS 39762
  (601) 325-5713
  Jack Coley

Mississippi Forestry Association
201 Realtors Building
620 North State Street
Jackson, MS 39201
  (601) 354-4936
  President: William Jones, Jr.

Mississippi
Forestry Commission
908 Robert E. Lee Building
Jackson, MS 39201
  (601) 354-7124
  State Forester: Billy Gaddis

Mississippi
Game and Fish Commission
Robert E. Lee Building
239 North Lamar Street
P.O. Box 451
Jackson, MS 39205
  (601) 354-7333
  Director of Conservation: Joe Stone

Mississippi Wildlife Federation
P.O. Box 1814
Jackson, MS 39205
  (601) 353-6922
  President: Edward Sullivan

Missouri
Department of Agriculture
Plant Industries Division
Jefferson Building

Jefferson City, MO 65101
  (314) 751-4211
  Director: Lester Barrows

Montana
Department of Agriculture
1300 Cedar Street
Helena, MT 59601
  (406) 449-3144
  Director: Gordon McOmber

Montana
Department of Agriculture
Pesticide Division
1300 Cedar Street
Helena, MT 59601
  (406) 449-3144
  Administrator: Gary Gingery

Montana
Department of Fish and Game
1420 East Sixth
Helena, MT 59601
  (406) 449-2535
  Director: Robert Wambach

Montana
Department of Fish and Game
Ecological Services Division
1420 East Sixth
Helena, MT 59601
  (406) 449-2335
  Administrator: James Posewitz

Montana
Department of Natural Resources and
  Conservation
Forestry Division
2705 Spurgin Road
Missoula, MT 59801
  (406) 728-4300
  Administrator: Gareth Moon

Montana
Environmental Quality Council
Capitol Station
Helena, MT 59601
  (406) 449-3742
    Ecology Researcher: William
      Harbrecht

Montana Wildlife Federation
Box 4373
Missoula, MT 59806
  (406) 549-2179
  President: Gary Stuker

National Association of State Depart-
  ments of Agriculture
Suite 710
1616 H Street, N.W.
Washington, DC 20006
  (202) 628-1566
  President: James Graham

National Association of State
  Foresters
Division of Forestry
Collins Building
Tallahassee, FL 32304
  President: John Bethea

National Audubon Society
950 Third Avenue
New York, NY 10022
  (212) 832-3200
  President: Elvis Stahr

National Waterfowl Council
Game and Fish Department
2222 West Greenway Road
Phoenix, AZ 85023
  (602) 942-3000
  Chairman: Robert Jantzen

Nebraska
Department of Agriculture
301 Centennial Mall South
P.O. Box 94947
Lincoln, NE 68509
  (402) 471-2341
  Director: Roger Sandman

Nebraska
Department of Agriculture
Bureau of Plant Industries
State Capitol
Lincoln, NE 68509
  (402) 471-2341
  Chief: Lloyd Bell

Nebraska
Department of Environmental Control
Agricultural Pollution Control
301 Centennial Mall South
P.O. Box 94947
Lincoln, NE 68509
  (402) 471-2341
  Chief: Jack Sukovaty

Nebraska
Game and Parks Commission
2200 North 33rd Street
P.O. Box 30370
Lincoln, NE 68503
  (402) 464-0641
  Director: Eugene Mahoney

Nebraska Ornithologists' Union
University of Nebraska State Museum
Lincoln, NE 68508
  President: Dr. Harvey Gunderson

Nebraska Wildlife Federation
R.R. 1
Box 122
Firth, NE 68358
  (402) 798-7406
  President: Francis Moul

Nevada
Department of Agriculture
350 Capitol Hill Avenue
P.O. Box 11100
Reno, NV 89510
  (702) 784-6401
  Executive Director: Thomas Ballow

Nevada
Department of Agriculture
Division of Plant Industry
350 Capitol Hill Avenue
Reno, NV 89510
  (702) 784-6401
  Phillip Martinelli

Nevada
Department of Conservation and
  Natural Resources
Division of Forestry
Capital Complex
201 South Fall

Carson City, NV 89710
  (702) 885-4350
  State Forester: Lowell Smith

Nevada
Department of Fish and Game
Box 10678
Reno, NV 89520
  (702) 784-6214
  Director: Glen Griffith

Nevada Wildlife Federation
P.O. Box 8022
University Station
Reno, NV 89507
  (702) 825-7823
  President: Eric Cronkhite

New Hampshire
Pesticide Control Division
State House Annex
Concord, NH 03301
  (603) 271-3550
  Control Supervisor: Murray McKay

New Hampshire Wildlife Federation
116 Hazelton Avenue
Manchester, NH 03103
  (603) 623-1885
  President: Joseph Ezyk

New Jersey Audubon Society
790 Ewing Avenue
Franklin Lakes, NJ 07417
  (201) 891-1211
  President: John Courtney

New Jersey
Department of Agriculture
P.O. Box 1888
Trenton, NJ 08625
  (609) 292-3976
  Secretary: Phillip Alampi

New Jersey
Department of Environmental Conser-
  vation
Division of Parks and Forestry
P.O. Box 1420
Trenton, NJ 08625
  (609) 292-2520
  State Forester: Gordon Bamford

New Jersey
Department of Environmental
  Protection
Office of Pesticide Control
P.O. Box 2807
Trenton, NJ 08625
  (609) 292-5890
  Chief: George Beyer

New Mexico
Department of Agriculture
Division of Pesticide Control
P.O. Box 3AQ
Las Cruces, NM 88003
  (505) 646-2133
  Chief: Barry Patterson

New Mexico Wildlife Federation
300 Val Verde, S.E.
Alburquerque, NM 87108
  (505) 265-7372
  President: Lloyd Haun, Jr.

New York
Cooperative Fishery Research Unit
Fernow Hall
Cornell University
Ithaca, NY 14853
  (607) 256-2151
  Leader: Dr. John Nickum

New York
Cooperative Wildlife Research Unit
Department of Natural Resources
Fernow Hall
Cornell University
Ithaca, NY 14853
  (607) 256-2014
  Leader: Dr. Milo Richmond

New York
Department of Agriculture and Markets
Building 8
State Campus
Albany, NY 12235
  (518) 457-2747
  Commissioner: Roger Barber

New York
Department of Environmental
  Conservation
Bureau of Pesticide Control

50 Wolf Road
Albany, NY 12233
  (518) 457-7482
  Director: Charles Frommer

New York Forest Owners Association
9 Grand Street
Cobleskill, NY 12043
  (518) 234-3813
  Secretary: Lewis Du Mond

New York
State Fish and Wildlife Management
  Board
R.D. 1
Rushville, NY 14544
  (315) 554-6510
  Chairman: Wilfred Kennedy

North Carolina
Cooperative Fishery Research Unit
Box 5577
4105 Gardner Hall
North Carolina State University
Raleigh, NC 27650
  (919) 755-4320
  Leader: Melvin Huish

North Carolina
Department of Agriculture
P.O. Box 27647
Raleigh, NC 27611
  (919) 733-7125
  Commissioner: James Graham

North Carolina
Department of Agriculture
Pesticide Control Division
P.O. Box 27647
Raleigh, NC 27611
  (919) 733-3556
  Administrator: William Buffaloe

North Carolina
Department of Natural Resources and
  Community Development
Forest Resources
P.O. Box 27687
Raleigh, NC 27611
  (919) 733-4984
  Ralph Winkworth

North Carolina
Department of Natural Resources and
    Community Development
Wildlife Resources Commission
P.O. Box 27687
Raleigh, NC 27611
    (919) 733-4984
    Executive Director: Bob Hazel

North Carolina Forestry Association
Box 19104
Raleigh, NC 27609
    (919) 834-3943
    Secretary: Ben Park

North Carolina Wildlife Federation
P.O. Box 10626
Raleigh, NC 27605
    (919) 782-5418
    President: Bryan Upchurch

North Dakota
Department of Agriculture
State Capitol
Bismark, ND 58501
    (701) 224-2374
    Pesticide Coordinator: Glen Johnson

North Dakota Wildlife Federation
R.R. 1
Carufel Addition
Bismark, ND 58501
    (702) 223-8741
    President: Calvin Helm

Northeast Association of Fish and
    Wildlife Resource Agencies
Division of Fish and Wildlife
Washington County Government Center
Tower Hill Road
Wakefield, RI 02878
    (401) 789-3094
    President: John Cronan

Northeastern Loggers Association
Information Office
Eagle Bay Road
Old Forge, NY 13420
    (315) 369-3078
    Executive Secretary

Pacific Fishery Management Council
526 S.W. Mill Street
Portland, OR 97201
    (503) 221-6352
    Executive Director: Lorry Nakatsu

Pacific Seabird Group
P.O. Box 1287
Juneau, AK 94802
    (916) 752-2108
    Chairman: Daniel Anderson

Pennsylvania
Department of Agriculture
Bureau of Plant Industry
2301 North Cameron Street
Harrisburg, PA 17120
    (717) 787-4843
    Lyle Forer

Pennsylvania Federation of Sportsmen's
    Clubs
1022 McCarter Avenue
Erie, PA 16503
    (814) 455-4032
    President: James Price

Pennsylvania Forestry Association
5221 East Simpson Street
Mechanicsburg, PA 17055
    President: Samuel Cobb

Ohio Audubon Council
4036 Cypress Road, N.E.
Canton, OH 44705
    (614) 486-4517
    President: Frank Starr, Jr.

Ohio Biological Survey
980 Biological Sciences Building
Ohio State University
484 West 12th Avenue
Columbus, OH 43210
    (614) 422-9645
    Executive Director: Dr. Charles
        King

Ohio
Department of Agriculture
Reynoldsburg, OH 43068
    (614) 466-2732
    Director: John Stackhouse

Ohio Forestry Association
665 East Dublin-Granville Road
Suite 205
Columbus, OH 43229
　(614) 846-9456
　Executive Director: Paul Bokros

Oklahoma
Department of Wildlife Conservation
1801 North Lincoln
P.O. Box 53465
Oklahoma City, OK 73105
　(405) 521-3851
　Environmental Coordinator: Ric
　Gomez

Oklahoma Ornithological Society
S.E.O.S.U.
Biology Department
Durant, OK 74701
　(405) 924-0121
　President: Constance Taylor

Oklahoma
State Department of Agriculture
Plant Industry Division
312 N.E. 28th Street
Oklahoma City, OK 73105
　(405) 521-3871
　Director: Dale Laubach

Oklahoma Wildlife Federation
Box 1262
Norman, OK 73070
　(405) 364-3609
　President: Dr. Thomas Peace

Oregon
Department of Agriculture
Plant Division
635 Capitol Street, N.E.
Salem, OR 97301
　(503) 378-3776
　Assistant Chief: Bill Kosesan

Oregon Wildlife Federation
P.O. Box 4552
Portland, OR 97208
　(503) 659-1457
　President: Larry Sowa

Rhode Island
Department of Environmental
　Management
Division of Agriculture
83 Park Street
Providence, RI 02903
　(401) 277-2781
　Chief: Rudolph D'Andrea

Society for the Protection of New
　Hampshire Forests
5 South State Street
Concord, NH 03301
　(603) 224-9945
　Chairman: Richard Webb

Society of American Foresters
5400 Grosvenor Lane
Washington, DC 20014
　(301) 897-8720
　President: Bernard Orell

South Carolina
Division of Regulatory and Public
　Service Programs
Barre Hall
Clemson Hall
Clemson University
Clemson, SC 29631
　(803) 656-3006
　Director: Dr. L.H. Senn

South Carolina Forestry Association
Suite 390
Dutch Center
Columbia, SC 29210
　(803) 798-4170
　President: Harold Kearse

South Carolina Wildlife Federation
Arcadian Plaza
Suite B-1
4949 Two Notch Road
Columbia, SC 29204
　(803) 786-6419
　President: Jasper Boles

South Dakota
Cooperative Fishery Research Unit
Department of Wildlife and Fisheries

South Dakota State University
Brookings, SD 57007
  (605) 688-6121
  Leader: Dr. Richard Applegate

South Dakota
Cooperative Wildlife Research Unit
South Dakota State University
Brookings, SD 57007
  (605) 688-6121
  Leader: Dr. Raymond Linder

South Dakota
Department of Agriculture
Sigurd Anderson Building
Pierre, SD 57501
  (605) 773-3375
  Secretary: Bob Duxbury

South Dakota
Department of Agriculture
Division of Agriculture Regulation
  and Inspection
Anderson Building
Pierre, SD 57501
  (605) 224-3351
  Acting Secretary

South Dakota Wildlife Federation
812 North Monroe
Pierre, SD 57501
  (605) 224-5360
  President: Roger Pries

South Dakota
Wildlife, Parks and Forestry Depart-
  ment
Forestry Division
Sigurd Anderson Building
Pierre, SD 57501
  (605) 224-3623
  Director: James Verville

South Dakota
Wildlife, Parks and Forestry Depart-
  ment
Wildlife Division
Sigurd Anderson Building
Pierre, SD 57501
  (605) 773-3381
  Director: Jerry Lounsberry

Southeastern Association of Fish and
  Wildlife Agencies
Department of Wildlife and Fisheries
400 Royal Street
New Orleans, LA 70130
  (504) 568-5667
  President: Burton Angelle

Southeastern Fishes Council
Department of Zoology
University of Tennessee
Knoxville, TN 37916
  (615) 974-2371
  Chairman: David Etnier

Southern Forest Institute
3395 Northeast Expressway
Suite 380
Atlanta, GA 30341
  (404) 451-7106
  President: W.H. Patterson

Tennessee
Department of Agriculture
Plant Industries Division
P.O. Box 40627
Nashville, TN 37204
  (615) 741-1551
  Assistant Director: Jim White

Tennessee Forestry Association
1720 West End Avenue
Nashville, TN 37203
  (615) 327-2473
  President: Richard Campbell

Texas
Department of Agriculture
Agricultural and Environmental
  Sciences Division
P.O. Box 12847
Capitol Station
Austin, TX 78711
  (512) 475-6346
  Director: David Ivie

Texas Forestry Association
P.O. Box 1488
Lufkin, TX 75901
  (713) 634-5523
  President: Rip Byrd

Texas
Forest Service
Forest Environmental Department
College Station, TX 77843
  (713) 845-2641
  Head: Mason Cloud, Jr.

Texas Organization for Endangered
  Species
P.O. Box 12773
Austin, TX 78711
  President: Dr. Clark Hubbs

Texas
Parks and Wildlife Department
Fisheries
4200 Smith School Road
Austin, TX 78744
  (512) 475-4888
  Director: Robert Kemp, Jr.

Texas
Parks and Wildlife Department
Wildlife
4200 Smith School Road
Austin, TX 78744
  (512) 475-4888
  Director: Ted Clark

U.S.
Department of Agriculture
Conservation, Research and Education
14th Street and Jefferson Drive, S.W.
Washington, DC 20250
  (202) 655-4000
  Assistant Secretary: Rupert Cutler

U.S.
Department of Agriculture
Forest Service
Washington, DC 20250
  (202) 447-3957
  Chief: John McGuire

U.S.
Department of Agriculture
Forest Service
Washington, DC 20250
  (202) 447-3957
  Director of Information: Robert
  Lake

U.S.
Department of Agriculture
Forest Service
Eisenhower Consortium for Environmen-
  tal Forestry Research
240 West Prospect
Fort Collins, CO 80524

U.S.
Department of Agriculture
Forest Service
Forest Products Lab
North Walnut Street
Box 5130
Madison, WI 53705
  (608) 257-2243
  Environmental Coordinator: Rodney
  DeGroot

U.S.
Department of Agriculture
Forest Service
Intermountain Station
507 - 25th Street
Ogden, UT 84401
  (801) 399-6286
  Environmental Coordinator: Jerry
  Sesco

U.S.
Department of Agriculture
Forest Service
North Central Area
1992 Folwell Avenue
St. Paul, MN 55108
  (612) 645-0251
  Environmental Coordinator: Roger
  Leonard

U.S.
Department of Agriculture
Forest Service
Northeastern Station
370 Reed Road
Upper Darby, PA 19008
  (215) 596-1615
  Environmental Coordinator: Ron
  Glass

U.S.
Department of Agriculture

Forest Service
Pacific Northwest Station
809 N.E. 6th Avenue
Box 3141
Portland, OR 97208
(503) 234-2059
Environmental Coordinator: Eldon
Estep

U.S.
Department of Agriculture
Forest Service
Pacific Southwest Station
1960 Addison Street
Box 245
Berkeley, CA 94701
(415) 486-3286
Environmental Coordinator: Richard
Hubbard

U.S.
Department of Agriculture
Forest Service
Pinchot Institute for Conservation
Studies
370 Reed Road
Broomall, PA 19008
(215) 596-1616
John Gray

U.S.
Department of Agriculture
Forest Service
Region 1
Federal Building
Missoula, MT 59807
(406) 329-3316
Regional Forester: Robert Torheim

U.S.
Department of Agriculture
Forest Service
Region 1
Federal Building
Missoula, MT 59807
(406) 329-3353
Environmental Coordinator: Ray
Franks

U.S.
Department of Agriculture
Forest Service

Region 2
Rocky Mountains
1117 West 8th Avenue
Box 25127
Lakewood, CO 80225
(303) 234-3820
Environmental Coordinator: Vacant

U.S.
Department of Agriculture
Forest Service
Region 2
Rocky Mountains
1117 West 8th Avenue
Box 25127
Lakewood, CO 80225
(303) 234-3711
Regional Forester: Craig Rupp

U.S.
Department of Agriculture
Forest Service
Region 3
Southwestern
Federal Building
517 Gold Avenue, S.W.
Alburquerque, NM 87102
(505) 766-3630
Environmental Coordinator: Don
Renton

U.S.
Department of Agriculture
Forest Service
Region 3
Federal Building
517 Gold Avenue, S.W.
Alburquerque, NM 87102
(505) 766-2401
Regional Forester: Milo Hassell

U.S.
Department of Agriculture
Forest Service
Region 4
Intermountain
Federal Building
324 25th Street
Ogden, UT 84401
(801) 399-6502
Environmental Coordinator: Mike
Griswold

U.S.
Department of Agriculture
Forest Service
Region 5
Pacific Southwest
630 Sansome Street
San Francisco, CA   94111
    (415) 556-3379
    Environmental Coordinator: George
    Coombes

U.S.
Department of Agriculture
Forest Service
Region 6
Pacific Northwest
319 S.W. Pine Street
Box 2623
Portland, OR   97208
    (503) 221-3865
    Environmental Coordinator: Curt
    Swanson

U.S.
Department of Agriculture
Forest Service
Region 8
Southern
1720 Peachtree Road, N.W.
Atlanta, GA   30309
    (404) 881-2242
    Environmental Coordinator: Jean
    Kruglewicz

U.S.
Department of Agriculture
Forest Service
Region 9
Eastern
633 West Wisconsin Avenue
Milwaukee, WI   53203
    (414) 224-3661
    Environmental Coordinator: Tom
    Hubbard

U.S.
Department of Agriculture
Forest Service
Region 10
Alaska
Federal Office Building
Box 1628

Juneau, AK
    (907) 586-7516
    Environmental Coordinator: Ray
    Clark

U.S.
Department of Agriculture
Forest Service
Rocky Mountain Station
240 West Prospect Street
Fort Collins, CO   80526
    (303) 221-1293
    Environmental Coordinator: Sam
    Krammes

U.S.
Department of Agriculture
Forest Service
Southeastern Station
Box 2570
Asheville, NC   28802
    (704) 258-0758
    Environmental Coordinator: Walter
    Hough

U.S.
Department of Agriculture
Forest Service
Southern Station
U.S. Postal Service Building
701 Loyola Avenue
New Orleans, LA   70113
    (504) 489-6712
    Environmental Coordinator: John
    Henley

U.S.
Department of Agriculture
Forest Service
Washington Office
P.O. Box 2417
Washington, DC   20013
    (202) 447-4710
    Environmental Coordinator: Dave
    Ketcham

U.S.
Department of Agriculture
Forest Service
Wildlife Service
Washington, DC   20250

(703) 235-8015
Director: Dale Jones

U.S.
Department of Agriculture
Office of Environmental Quality
Room 412-A
Administration Building
Washington, DC 20250
    (202) 447-3965
    Director: Barry Flamm

U.S.
Department of Agriculture
Soil Conservation Service
Alabama
P.O. Box 311
Auburn, AL 36830
    Field Biologist: Robert Waters

U.S.
Department of Agriculture
Soil Conservation Service
Arizona
6029 Federal Building
230 North First Avenue
Phoenix, AZ 85025
    Field Biologist: John York

U.S.
Department of Agriculture
Soil Conservation Service
Arkansas
P.O. Box 2323
Little Rock, AR 72203
    Field Biologist: Paul Brady

U.S.
Department of Agriculture
Soil Conservation Service
California
P.O. Box 1019
Davis, CA 95616
    Field Biologist: Wendell Miller

U.S.
Department of Agriculture
Soil Conservation Service
Colorado
2490 West 26th Avenue
Room 309
Denver, CO 80211

Field Biologist: Eldie Mustard, Jr.

U.S.
Department of Agriculture
Soil Conservation Service
District of Columbia
P.O. Box 2890
Washington, DC 20013
    (202) 447-5991
    Chief Biologist: Carl Thomas

U.S.
Department of Agriculture
Soil Conservation Service
Florida
P.O. Box 1208
Gainesville, FL 32602
    Field Biologist: John Vance

U.S.
Department of Agriculture
Soil Conservation Service
Georgia
P.O. Box 832
Athens, GA 30601
    Field Biologist: Jesse Mercer

U.S.
Department of Agriculture
Soil Conservation Service
Idaho
304 North 8th Street
Room 345
Boise, ID 83702
    Field Biologist: Clyde Scott

U.S.
Department of Agriculture
Soil Conservation Service
Illinois
P.O. Box 678
Champaign, IL 61820
    Field Biologist: Rex Hamilton

U.S.
Department of Agriculture
Soil Conservation Service
Indiana
5610 Crawfordsville Road
Indianapolis, IN 46224
    Field Biologist: James McCall

U.S.
Department of Agriculture
Soil Conservation Service
Iowa
823 Federal Office Building
210 Walnut Street
Des Moines, IA 50309
    Field Biologist: Lyle Asel

U.S.
Department of Agriculture
Soil Conservation Service
Kansas
P.O. Box 600
Salina, KS 67401
    Field Biologist: Richard Hager

U.S.
Department of Agriculture
Soil Conservation Service
Kentucky
333 Waller Avenue
Lexington, KY 40504
    Field Biologist: William Casey

U.S.
Department of Agriculture
Soil Conservation Service
Louisiana
3737 Government Street
P.O. Box 1630
Alexandria, LA 71301
    Field Biologist: Edward Smith

U.S.
Department of Agriculture
Soil Conservation Service
Maine
USDA Office Building
University of Maine
Orono, ME 04473
    Field Biologist: Robert Wengrzynek

U.S.
Department of Agriculture
Soil Conservation Service
Maryland
4321 Hartwick Road
Room 522
College Park, MD 20740
    Field Biologist: Eugene Whitaker

U.S.
Department of Agriculture
Soil Conservation Service
Michigan
1405 South Harrison Road
Room 101
East Lansing, MI 48823
    Field Biologist: Charles Smith

U.S.
Department of Agriculture
Soil Conservation Service
Minnesota
316 North Robert Street
St. Paul, MN 55101
    Field Biologist: Allen Vaughn

U.S.
Department of Agriculture
Soil Conservation Service
Mississippi
P.O. Box 610
Jackson, MS 39205
    Field Biologist: Edward Sullivan

U.S.
Department of Agriculture
Soil Conservation Service
Missouri
P.O. Box 459
Columbia, MO 65201
    Field Biologist: Edward Gaskins

U.S.
Department of Agriculture
Soil Conservation Service
Montana
P.O. Box 970
Bozeman, MT 59715
    Field Biologist: Ronald Batchelor

U.S.
Department of Agriculture
Soil Conservation Service
Nebraska
Room 604
134 South 12th Street
Lincoln, NE 68508
    Field Biologist: Robert Koerner

U.S.
Department of Agriculture

Soil Conservation Service
Nebraska
100 Centennial Mall North
Lincoln, NE 68508
  Regional Biologist: Gerald Lowry

U.S.
Department of Agriculture
Soil Conservation Service
Nevada and Utah
125 South State Street
Salt Lake City, UT 84111
  Field Biologist: David Chalk

U.S.
Department of Agriculture
Soil Conservation Service
New Hampshire
Federal Building
P.O. Box G
Durham, NH 03824
  Field Biologist: David Allen

U.S.
Department of Agriculture
Soil Conservation Service
New Jersey
1370 Hamilton Street
P.O. Box 219
Somerset, NJ 08873
  Field Biologist: Dave Smart

U.S.
Department of Agriculture
Soil Conservation Service
New Mexico
P.O. Box 2007
Albuquerque, NM 87103
  Field Biologist: Edwin Senson, Jr.

U.S.
Department of Agriculture
Soil Conservation Service
New York
700 East Water Street
Syracuse, NY 13210
  Field Biologist: Francis Keeler

U.S.
Department of Agriculture
Soil Conservation Service
North Carolina

P.O. Box 27307
Raleigh, NC 27611
  Field Biologist: John Edwards

U.S.
Department of Agriculture
Soil Conservation Service
North Dakota
P.O. Box 1458
Bismark, ND 58501
  Field Biologist: Erling Podoll

U.S.
Department of Agriculture
Soil Conservation Service
Ohio
311 Old Federal Building
Columbus, OH 43215
  Field Biologist: Dennis Haag

U.S.
Department of Agriculture
Soil Conservation Service
Oregon
Federal Building
511 N.W. Broadway
Portland, OR 97209
  Regional Biologist: Dean Marriage

U.S.
Department of Agriculture
Soil Conservation Service
Oregon
1218 S.W. Washington
Portland, OR 97205
  Field Biologist: Robert Corthell

U.S.
Department of Agriculture
Soil Conservation Service
Pennsylvania
Box 985
Federal Square Station
Harrisburg, PA 17108
  Field Biologist: Clayton Heiney, Jr.

U.S.
Department of Agriculture
Soil Conservation Service
Pennsylvania
1974 Sproul Road
Broomhall, PA 19008

Regional Biologist: Vernon Hicks

U.S.
Department of Agriculture
Soil Conservation Service
South Carolina
240 Stoneridge Drive
Columbia, SC 29201
   Field Biologist: William Melven

U.S.
Department of Agriculture
Soil Conservation Service
South Dakota
P.O. Box 1357
Huron, SD 57350
   Field Biologist: John Farley

U.S.
Department of Agriculture
Soil Conservation Service
Tennessee
561 U.S. Courthouse
Nashville, TN 37203
   Field Biologist: Floyd Fessler

U.S.
Department of Agriculture
Soil Conservation Service
Texas
P.O. Box 648
Temple, TX 76501
   Field Biologist: James Henson

U.S.
Department of Agriculture
Soil Conservation Service
Texas
P.O. Box 6567
Fort Worth, TX 76115
   Regional Biologist: Edward Smith

U.S.
Department of Agriculture
Soil Conservation Service
Utah
125 State Street
Salt Lake City, UT 84111
   Field Biologist: David Clark

U.S.
Department of Agriculture

Soil Conservation Service
Virginia
P.O. Box 10026
Richmond, VA 23240
   Field Biologist: Franklin Dugan

U.S.
Department of Agriculture
Soil Conservation Service
Washington
West 920 Riverside Avenue
Spokane, WA 99201
   Field Biologist: Ivan Hines, Jr.

U.S.
Department of Agriculture
Soil Conservation Service
West Virginia
P.O. Box 865
Morgantown, WV 26505
   Field Biologist: Thomas Crebbs

U.S.
Department of Agriculture
Soil Conservation Service
Wisconsin
P.O. Box 4248
Madison, WI 53711
   Field Biologist: Steve Baima

U.S.
Department of Agriculture
Soil Conservation Service
Wyoming
P.O. Box 2440
Casper, WY 86202
   Field Biologist

U.S.
Department of Commerce
NOAA
National Marine Fisheries Service
Washington, DC 20235
   (202) 634-7283
   Assistant Administrator for Fish-
      eries: Terry Leitzell

U.S.
Department of Commerce
NOAA
National Marine Fisheries Service
Alaska Region

Federal Building
Room 453
709 West Ninth Street
P.O. Box 1668
Juneau, AK 99802
  (907) 586-7221
  Director: Vacant

U.S.
Department of Commerce
NOAA
National Marine Fisheries Service
Atlantic Environmental Group
Route 7A
P.O. Box 522A
Narragansett, RI 02882
  (401) 789-9326
  Director: Merton Ingham

U.S.
Department of Commerce
NOAA
National Marine Fisheries Service
Environmental and Technical Services
  Division
811 N.E. Oregon Street
P.O. Box 4332
Portland, OR 97208
  (503) 234-3361
  Chief: Dale Evans

U.S.
Department of Commerce
NOAA
National Marine Fisheries Service
Northeast Region
Federal Building
14 Elm Street
Gloucester, MA 01930
  (617) 281-3600
  Director: Allen Peterson, Jr.

U.S.
Department of Commerce
NOAA
National Marine Fisheries Service
Northwest Region
1700 Westlake Avenue, North
Seattle, WA 98109
  (206) 442-7575
  Director: Vacant

U.S.
Department of Commerce
NOAA
National Marine Fisheries Service
Office of Habitat Protection
Washington, DC 20235
  (202) 634-7490
  Director: James Rote

U.S.
Department of Commerce
NOAA
National Marine Fisheries Service
Office of Marine Mammals and Endan-
  gered Species
Washington, DC 20235
  (202) 634-7461
  Director: William Aron

U.S.
Department of Commerce
NOAA
National Marine Fisheries Service
Office of Policy and Planning
Washington, DC 20235
  (202) 634-7430
  Director: Richard Gutting, Jr.

U.S.
Department of Commerce
NOAA
National Marine Fisheries Service
Office of Resource Conservation and
  Management
Washington, DC 20235
  (202) 634-7218
  Director: William Gordon

U.S.
Department of Commerce
NOAA
National Marine Fisheries Service
Office of Science and Environment
Washington, DC 20235
  (202) 634-7469
  Director: Vacant

U.S.
Department of Commerce
NOAA
National Marine Fisheries Service

Pacific Environmental Group
c/o Fleet Numerical Weather Central
Monterey, CA 93940
    (408) 373-3331
    Chief: Gunther Seckel

U.S.
Department of Commerce
NOAA
National Marine Fisheries Service
Southeast Region
Duval Building
9450 Koger Boulevard
St. Petersburg, FL 33702
    (813) 893-3142
    Director: William Stevenson

U.S.
Department of Commerce
NOAA
National Marine Fisheries Service
Southwest Region
300 South Ferry Street
Terminal Island, CA 90731
    (213) 548-2575
    Director: Vacant

U.S.
Department of the Interior
Conservation and Wildlife
C Street Between 18th and 19th
    Streets, N.W.
Washington, DC 20240
    (202) 343-1100
    Associate Solicitor: James Webb

U.S.
Department of the Interior
Endangered Species Commission
Room 4160
Washington, DC 20240
    (202) 343-5978

U.S.
Department of the Interior
Engineering Office
18th and C Streets, N.W.
Washington, DC 20240
    (202) 343-3193

U.S.
Department of the Interior

Fish and Wildlife Service
Washington, DC 20240
    (202) 343-4717
    Associate Director: Eugene Hester

U.S.
Department of the Interior
Fish and Wildlife Service
Albuquerque Regional Office
P.O. Box 1306
Albuquerque, NM 87103
    (505) 766-2321

U.S.
Department of the Interior
Fish and Wildlife Service
Anchorage Regional Office
813 D Street
Anchorage, AK 99501
    (907) 276-3800

U.S.
Department of the Interior
Fish and Wildlife Service
Atlanta Regional Office
17 Executive Park Drive, N.E.
Atlanta, GA 30329
    (404) 881-4671

U.S.
Department of the Interior
Fish and Wildlife Service
Biological Services Office
1730 K Street, N.W.
Washington, DC 20240
    (202) 634-4910

U.S.
Department of the Interior
Fish and Wildlife Service
Boston Regional Office
One Gateway Center
Suite 700
Newton Corner, MA 02158

U.S.
Department of the Interior
Fish and Wildlife Service
Denver Regional Office
P.O. Box 25486
Denver Federal Center
Denver, CO 80225

(303) 234-2209

U.S.
Department of the Interior
Fish and Wildlife Service
Ecological Services Division
711 14th Street, N.W.
Washington, DC 20240
  (202) 376-8121
  Dr. Alfred Fox

U.S.
Department of the Interior
Fish and Wildlife Service
Habitat Preservation Research
  Division
18th and C Streets, N.W.
Washington, DC 20240
  (202) 343-7557

U.S.
Department of the Interior
Fish and Wildlife Service
Information Transfer Network
Washington, DC 20240
  (202) 343-8032

U.S.
Department of the Interior
Fish and Wildlife Service
Law Enforcement
P.O. Box 19183
Washington, DC 20236
  (202) 343-9242

U.S.
Department of the Interior
Fish and Wildlife Service
Migratory Bird Management
18th and C Streets, N.W.
Washington, DC 20240
  (202) 254-3207

U.S.
Department of the Interior
Fish and Wildlife Service
National Coastal Ecosystems Team
National Space Technology Laboratory
NSTL Station, MS 39529
  (601) 688-2091

U.S.
Department of the Interior
Fish and Wildlife Service
National Laboratory
10th Street and Constitution Avenue,
  N.W.
Washington, DC 20240
  (202) 381-5161

U.S.
Department of the Interior
Fish and Wildlife Service
National Reservoir Research Program
Fayetteville, AR 72701
  (501) 521-3063
  Director

U.S.
Department of the Interior
Fish and Wildlife Service
National Wildlife Refuges Division
18th and C Streets, N.W.
Washington, DC 20240
  (202) 343-4791

U.S.
Department of the Interior
Fish and Wildlife Service
Northern Prairie Wildlife Research
  Center
Jamestown, ND 58401
  (701) 252-5363

U.S.
Department of the Interior
Fish and Wildlife Service
Patuxent Wildlife Research Center
Laurel, MD 20811
  (301) 776-4880

U.S.
Department of the Interior
Fish and Wildlife Service
Portland Regional Office
P.O. Box 3737
Portland, OR 97208
  (503) 231-6118

U.S.
Department of the Interior

Fish and Wildlife Service
Twin Cities Regional Office
Federal Building
Fort Snelling
Twin Cities, MN   55111
   (612) 725-3500

U.S.
Department of the Interior
Fish and Wildlife Service
Western Energy and Landuse Team
2625 Redwing Road
Fort Collins, CO   80526
   (303) 221-2040
   Information Transfer Specialist

U.S.
Department of the Interior
Fish and Wildlife Service
Wildlife Research Division
1717 H Street, N.W.
Washington, DC   20240
   (202) 343-7557

U.S.
Department of the Interior
Migratory Bird Conservation
   Commission
Department of the Interior Building
Washington, DC   20240
   (202) 343-4676

U.S.
Endangered Species Scientific
   Authority
1717 H Street, N.W.
Washington, DC   20555
   (202) 653-5948

U.S.
Environmental Protection Agency
Environmental Impact Review Staff
Region V
230 Dearborn Street
Chicago, IL   60604
   (312) 353-2000
   Aquatic Biologist: James Hooper

U.S.
Environmental Protection Agency
EIS Review Section
Region IV

345 Courtland Street
Atlanta, GA   30308
   (404) 881-4727
   Environmental Specialist: Gerald
   Miller

U.S.
Marine Mammal Commission
Room 307
1625 I Street, N.W.
Washington, DC   20006
   (202) 653-6237

U.S.
Tennessee Valley Authority
Forestry, Fisheries, and Wildlife
   Development
400 Commerce Avenue
Knoxville, TN   37902
   (615) 632-2101
   Director: Thomas Ripley

Urban Wildlife Research Center
12789 Folly Quarter Road
Ellicott City, MD   21043
   (301) 596-5553
   President: Dr. Guttermuth

Utah
Cooperative Fishery Unit
Utah State University
UMC 52
Logan, UT   84322
   (801) 752-4100
   Leader: Dr. Charles Berry, Jr.

Utah
Cooperative Wildlife Research Unit
Utah State University
Logan, UT   84322
   (801) 752-4100
   Leader: Dr. David Anderson

Utah
Department of Agriculture
147 North 200 West
Salt Lake City, UT   84103
   (801) 533-5421
   Commissioner: Kenneth Creer

Utah
Department of Agriculture

Division of Plant Industry
147 North 200 West
Salt Lake City, UT 84114
   (801) 533-4107
   Director: Wanless Southwick

Utah
Department of Agriculture
Environmental Coordinating Committee
124 State Capitol
Salt Lake City, UT 84111
   (801) 533-5794
   Carolyn Lloyd

Utah
Department of Agriculture
Environmental Coordinating Committee
Division of Wildlife Resources
124 State Capitol
Salt Lake City, UT 84111
   (801) 533-5794
   Earl Sparks

Utah
State Department of Natural Resources
Division of Wildlife Resources
1596 W.N. Temple
Salt Lake City, UT 84116
   (801) 533-9333
   Director: Douglas Day

Utah
State Department of Natural Resources
Division of Wildlife Resources
Fisheries Management
1596 W.N. Temple
Salt Lake City, UT 84116
   (801) 533-9333
   Chief: Don Andriano

Utah
State Department of Natural Resources
Division of Wildlife Resources
Game Management
1596 W.N. Temple
Salt Lake City, UT 84116
   (801) 533-9333
   Chief: Norman Hancock

Utah
State Department of Natural Resources
Forestry and Fire Control

State Forester's Office
440 Empire Building
231 East 4th South
Salt Lake City, UT 84111
   (801) 533-5439
   State Forester: Paul Sjoblom

Vermont
Agency of Environmental Conservation
Fisheries Management
Montpelier, VT 05602
   (802) 828-3371
   Angelo Incerpi

Vermont
Agency of Environmental Conservation
Forestry Division
Montpelier, VT 05602
   (802) 828-3471
   Director: Bradford Walker

Vermont
Agency of Environmental Conservation
Game Management
Montpelier, VT 05602
   (802) 828-3371
   Benjamin Day

Vermont
Department of Agriculture
Division of Plant Pest Control
Montpelier, VT 05602
   (802) 826-2431
   Director: Phillip Benedict

Virginia
Commission of Game and Inland Fish-
   eries
4010 West Broad Street
P.O. Box 11104
Richmond, VA 23230
   (804) 257-1000
   Executive Director: James McInteer

Virginia
Commission of Game and Inland Fish-
   eries
Fish Division
4010 West Broad Street
Box 11104
Richmond, VA 23230
   (804) 257-1000

Chief: Jack Hoffman

Virginia
Commission of Game and Inland Fisheries
Game Division
4010 West Broad Street
Box 11104
Richmond, VA 23230
    (804) 257-1000
    Chief: Richard Cross, Jr.

Virginia
Cooperative Fishery Research Unit
Virginia Polytechnic Institute and
    State University
106 Cheatham Hall
Blacksburg, VA 24061
    (703) 951-5507
    Leader: Garland Pardue

Virginia
Cooperative Wildlife Research Unit
Virginia Polytechnic Institute and
    State University
100 Cheatham Hall
Blacksburg, VA 24061
    (703) 961-5573
    Leader: Dr. Burd McGinnes

Virginia
Department of Agriculture and
    Consumer Services
203 North Govenor Street
Room 405
Richmond, VA 23219
    (804) 786-3978
    Chief: Dr. Berkwood Framer

Virginia
Department of Agriculture and
    Consumer Services
Division of Product and Industrial
    Regulations
Pesticide, Paint, and Hazardous
    Substances Section
1204 East Main Street
Richmond, VA 23219
    (804) 786-3798
    Supervisor: Harry Rust

Virginia
Department of Conservation and Economic Development
Division of Forestry
Box 3758
Charlottesville, VA 22903
    State Forester: Wallace Custard

Virginia Forestry Association
One North Fifth Street
Richmond, VA 23219
    (804) 644-8462
    President: H.E. Matics

Virginia Society of Ornithology
Department of Biology
University of Virginia
Charlottesville, VA 22903
    (804) 924-7868
    President: J.J. Murray, Jr.

Virginia Wildlife Federation
P.O. Box 3609
Norfolk, VA 23514
    (804) 627-0055
    President: Walter Leveridge

Washington
Department of Agriculture
406 General Administration Building
Olympia, WA 98504
    (206) 753-5063
    Director: Bob Mickelson

Washington
Department of Fisheries
115 General Administration Building
Olympia, WA 98504
    (206) 753-6600
    Director: Gordon Sandison

Washington
Department of Game
Division of Environmental Management
600 North Capitol Way
Olympia, WA 98504
    (206) 753-5700
    Chief: Eugene Dziedzic

Washington
Department of Game

Division of Fishery Management
600 North Capitol Way
Olympia, WA 98504
  (206) 753-5700
  Chief: Cliff Millenbach

Washington
Department of Game
Division of Wildlife Management
600 North Capitol Way
Olympia, WA 98504
  (206) 753-5700
  Chief: Wallace Krammer

Washington Forest Protection
  Association
Suite 1220
1411 Fourth Avenue Building
Seattle, WA 98101
  (206) 623-1500
  President: Max Schmidt

Washington State Forestry Conference
711 Capitol Way
Suite 608
Olympia, WA 98501
  (206) 352-1500
  President: James Bethel

Western Association of State Game
  and Fish Commissioners
Idaho Fish and Game Department
Box 25
Boise, ID 83707
  (208) 384-3772
  Secretary: Robert Salter

Western Forestry and Conservation
  Association
1326 American Bank Building
Portland, OR 97205
  (503) 226-4562
  President: John Callaghan

West Virginia
Department of Agriculture
Plant Pest Control Division
Capitol Building
Charleston, WV 25305
  (304) 348-2201
  Director: Albert Cole

West Virginia Wildlife Federation
Box 275
Paden City, WV 26159
  (304) 337-9166
  President: Olston Wright

Wetlands for Wildlife
P.O. Box 147
Mayville, WI 53050
  (414) 281-8936
  President: Richard Goff

Wilderness Watch
Office of the President
P.O. Box 3184
Green Bay, WI 54303
  (414) 499-9131
  President: Dr. Jerry Gaudt

Wildfowl Foundation
709 Wire Building
1000 Vermont Avenue, N.W.
Washington, DC 20005
  (202) 347-1774
  President: C.R. Gutermuth

Wildlife Federation
Pennsylvania Chapter
RD 2
Boiling Springs, PA 17007
  (717) 787-6286
  President: Jerry Hassinger

Wildlife Management Institute
709 Wire Building
1000 Vermont Avenue, N.W.
Washington, DC 20005
  (202) 347-1774
  President: Daniel Poole

Wildlife Resources
Box 234
Troy, ID 83871
  (208) 835-3891
  President: Harvey Neese

Wildlife Society
Suite 611
7101 Wisconsin Avenue, N.W.
Washington, DC 20014
  (301) 986-8700
  President: Leslie Pengelly

Wildlife Society
Alabama Chapter
P.O. Box 352
Jackson, AL 36104
 (205) 246-2165
 President: James Davis

Wildlife Society
Alaska Chapter
20145-D
Fairbanks, AK 99701
 (902) 277-8548
 President: Richard Bishop, Sr.

Wildlife Society
Arizona Chapter
P.O. Box 35414
Phoenix, AZ 85069
 President: Donald Seibert

Wildlife Society
Arkansas Chapter
P.O. Box 391
Little Rock, AR 72203
 (501) 376-6301
 President: James Miller

Wildlife Society
Colorado Chapter
160 South Ames
Lakewood, CO 80226
 (303) 234-3699
 President: Farrell Copelin

Wildlife Society
Florida Chapter
U.S. Fish and Wildlife Service
2820 East University Avenue
Gainesville, FL 32601
 President: Dr. Nicholas Holler

Wildlife Society
Georgia Chapter
RT 1
Rutledge, GA 30663
 (404) 557-2706
 President: Ronald Odom

Wildlife Society
Hawaii Chapter
41-875 Laumilo Street

Waimanalo
Oahu, HI 96795
 President: David Woodside

Wildlife Society
Idaho Chapter
3507 Buckskin Road
Coeur d'Alene, ID 83814
 (208) 667-2561
 President: Dean Carrier

Wildlife Society
Illinois Chapter
Illinois Department of Conservation
P.O. Box 728
Hinckley, IL 60520
 President: George Hubert, Jr.

Wildlife Society
Indiana Chapter
607 State Office Building
Indianapolis, IN 46204
 (317) 633-7696
 President: Russell Hyer

Wildlife Society
Kansas Chapter
1803 West 6th
Emporia, KS 66801
 President: Bill Peabody

Wildlife Society
Maine Chapter
University of Maine
240 Nutting Hall
Orono, ME 04473
 (207) 581-7386
 President: Malcomb Coulter

Wildlife Society
Minnesota Chapter
Biology Department
St. Cloud State University
St. Cloud, MN 56301
 (612) 255-4135
 President: Alfred Grewe, Jr.

Wildlife Society
Mississippi Chapter
Box 610
Jackson, MS 39205

(601) 969-4339
President: Edward Sullivan

Wildlife Society
Missouri Chapter
1202 Sycamore Drive
Rola, MO 65401
    (314) 368-7749
    President: Gary Houf

Wildlife Society
Montana Chapter
Box 5
Montana State University
Bozeman, MT 59717
    (406) 994-3285
    President: John Wiegand

Wildlife Society
National Capitol Chapter
Fish and Wildlife Service
EELUG
Harper's Ferry Center
Harper's Ferry, WV 25425
    President: Charles Cushwa

Wildlife Society
Nevada Chapter
P.O. Box 1242
Tonopah, NV 89049
    President: Vacant

Wildlife Society
New Jersey Chapter
P.O. Box 9
Chester, NJ 07930
    (201) 735-8793
    President: Robert Lund

Wildlife Society
New Mexico Chapter
Department of Game and Fish
Sante Fe, NM 87501
    (505) 988-6588
    President: Keith Giezentanner

Wildlife Society
New York Chapter
P.O. Box 535
New Paltz, NY 12561
    (914) 692-6706
    President: Warren McKeon

Wildlife Society
North Dakota Chapter
2121 Lovett Avenue
Bismark, ND 58505
    President: Wilbur Boldt

Wildlife Society
Northern Virginia Chapter
13125 Pennerview Lane
Fairfax, VA 22030
    (202) 755-2972
    President: Jerry Moore

Wildlife Society
Ohio Chapter
8589 Horsehoe Road
Ashley, OH 43003
    (614) 747-2525
    President: John Harder

Wildlife Society
Oklahoma Chapter
1212 Greenbriar Court
Norman, OK 73069
    (405) 521-2730
    President: Byron Moser

Wildlife Society
Oregon Chapter
P.O. Box 3503
Portland, OR 97208
    (503) 229-5463
    President: James Harper

Wildlife Society
South Carolina Chapter
Department of Economic Zoology
Clemson University
Clemson, SC 29631
    (803) 656-3111
    President: John Sweeney

Wildlife Society
South Dakota Chapter
Wildlife and Fish Department
South Dakota State University
Brookings, SD 57007
    (605) 688-6121
    President: Lester Flake

Wildlife Society
Tennessee Chapter

Department of Biology
Memphis State University
Memphis, TN 38152
  (901) 454-2586
  President: Michael Kennedy

Wildlife Society
Texas Chapter
Department of Wildlife and Fisheries
  Sciences
Texas A&M University
College Station, TX 77843
  (713) 845-7471
  President: Charles Ramsey

Wildlife Society
Utah Chapter
Utah Division of Wildlife Research
1596 W.N. Temple
Salt Lake City, UT 84116
  (801) 533-9333
  President: Rodney John

Wildlife Society
Washington Chapter
5404 N.E. Hazel Dell Avenue
Vancouver, WA 98663
  (206) 696-6211
  President: Richard Poelker

Wildlife Society
West Virginia Chapter
USDA Forest Service
180 Canfield Street
Morgantown, WV 26505
  (304) 599-7487
  President: Vernon Henry

Wildlife Society
Wisconsin Chapter
RT 2
Box 150 C
Cambridge, WI 53523
  (608) 262-2671
  President: Donald Rusch

Wildlife Society
Wyoming Chapter
4404 East 8th Street
Cheyenne, WY 82001
  (307) 777-7738
  President: Don Johnson

Wilson Ornithological Society
Department of Zoology
University of Arkansas
Fayetteville, AR 72701
  (501) 575-3251
  President: Douglas James

Wisconsin
Cooperative Fishery Research Unit
College of Natural Resources
University of Wisconsin
Stevens Point, WI 54481
  (715) 346-2178
  Leader: Dr. Daniel Coble

Wisconsin
Cooperative Wildlife Research Unit
Department of Wildlife Ecology
226 Russell Laboratories
University of Wisconsin
Madison, WI 53706
  (608) 262-2671
  Leader: Dr. Donald Rusch

Wisconsin
Department of Agriculture
801 West Badger Road
Madison, WI. 53713
  (608) 266-7102
  Secretary: Dr. Gary Rohde

Wisconsin
Department of Natural Resources
Bureau of Fish Management
Box 7921
Madison, WI 53707
  (608) 266-7025
  Director: James Addis

Wisconsin
Department of Natural Resources
Bureau of Forestry
Box 7921
Madison, WI 53707
  (608) 266-0842
  Director: Milton Reinke

Wisconsin
Department of Natural Resources
Bureau of Wildlife Management
Box 7921
Madison, WI 53707

(608) 266-2193
Director: John Keener

<u>Wisconsin Society for Ornithology</u>
2 Pioneer Place
Elgin, WI 60120
  (312) 695-2464
  President: Daryl Tessen

<u>Wisconsin Wildlife Federation</u>
301½ Main Street
Mosinee, WI 54455
  (715) 692-2242
  President: Harold Spencer

<u>World Wildlife Fund</u>
1601 Connecticut Avenue, N.W.
Suite 800
Washington, DC 20009
  (202) 387-0805
  President: Russell Train

<u>Wyoming</u>
Department of Agriculture
Division of Plant Industry
2219 Cary Street
Cheyenne, WY 82002
  Director: Walter Patch

<u>Wyoming Wildlife Federation</u>
P.O. Box 826
Riverton, WY 82501
  (307) 856-5768
  President: Al Conrad

7. WATER-RELATED AGENCIES AND ORGANIZATIONS

Academy of Natural Sciences
Limnology Department
19th Street and the Parkway
Philadelphia, PA 19103
(215) 299-1112
Information Services

Alabama
Environmental Health Administration
Division of Public Water Supplies
State Office Building
Montgomery, AL 36130
(205) 832-3170
Director: Joseph Downey

Alabama
Water Improvement Commission
State Office Building
Montgomery, AL 36130
(205) 832-3370
Chief Administration Officer:
James Warr

Alaska
Department of Environmental Conser-
vation
Water Quality Management Section
Pouch O
Juneau, AK 99811
(907) 465-2643
Chief: Bob Martin

American National Standards Institute
ISO Technical Committee
147 Water Quality

1430 Broadway
New York, NY 10018
(212) 354-3346
Secretariat: Israel Resnick

American Rivers Conservation Council
317 Pennsylvania Avenue, S.E.
Washington, DC 20003
(202) 547-6900
Director: Howard Brown

American Society of Limnology and
Oceanography
Oregon State University
School of Oceanography
Corvallis, OR 97331
President: Robert Smith

American Water Resources Association
St. Anthony Falls Hydraulic Laboratory
Mississippi at Third Avenue, S.E.
Minneapolis, MN 55414
(612) 376-5050
President: Daniel Evans

American Water Works Association
Office of the Director
6666 West Quincy
Denver, CO 80235
(303) 794-7711
Executive Director

Arizona
Bureau of Water Quality Control
1740 West Adams Street

Phoenix, AZ   85007
  (602) 255-1252
  Chief: Dr. Ronald Miller

Arkansas
Department of Health
4815 West Markham Street
Little Rock, AR   72201
  (501) 661-2111
  Director: Robert Young

Arkansas
Department of Pollution Control and
  Ecology
208 Planning
8001 National Drive
Little Rock, AR   72209
  (501) 371-1701
  Chief: Johnie Brown

Arkansas
Department of Pollution Control and
  Ecology
Water Division
8001 National Drive
Little Rock, AR   72209
  (501) 371-1701
  Chief: Jim Shell

California
Department of Water Resources
1416 Ninth Street
Sacramento, CA   95814
  (916) 445-7416
  Ken Fellows

California
State Water Resources Control Board
2125 19th Street
Sacramento, CA   95814
  (916) 445-0847
  John Huddleson

Clean Water Action Project
1341 G Street, N.W.
Washington, DC   20005
  (202) 638-1196
  President: David Zwick

Colorado
Department of Health
Water Quality Board

4210 East 11th Avenue
Denver, CO   80220
  (303) 320-8333
  Section Chief: George Prince

Colorado Water Congress
1111 South Colorado Boulevard
Suite 401
Denver, CO   80222
  (303) 759-9805
  President: William Raley

Connecticut
Department of Environmental Protection
Water Resources Unit
State Office Building
165 Capitol Avenue
Hartford, CT   06115
  (203) 566-5599
  Director: Ben Warner

Connecticut
Department of Health
Water Supply Section
79 Elm Street
Hartford, CT   06115
  (203) 566-3130
  Chief: Richard Woodhull

Delaware
Department of Health and Social
  Services
Bureau of Environmental Health
Jesse Cooper Building
Capitol Square
Dover, DE   19901
  (302) 678-4731
  Chief: Donald Harmeson

Delaware
Department of Natural Resources and
  Environmental Control
Division of Environmental Control
Water Resources Section
Tatnall Building
P.O. Box 1401
Dover, DE   19901
  (302) 678-4764
  Manager: Robert Touhey

District of Columbia
Department of Environmental Services

Bureau of Air and Water Quality
5010 Overlook Avenue, S.W.
Washington, DC 20002
    (202) 767-7486
    Chief: John Brink

Florida
Department of Environmental Regula-
    tions
Division of Environmental Programs
2562 Executive Center Circle, East
Tallahassee, FL 32301
    (904) 487-1855
    Acting Director: William Townsend

Georgia
Department of Natural Resources
Environmental Protection Division
Water Protection Branch
270 Washington Street, S.W.
Room 822
Atlanta, GA 30334
    (404) 656-6593
    Chief: Gene Welsh

Georgia
Department of Natural Resources
Water Resources Survey
270 Washington Street, S.W.
Atlanta, GA 30334
    (404) 656-3530
    Chief: David Swanson

Hawaii
Department of Land and Natural
    Resources
Division of Water and Land Develop-
    ment
P.O. Box 373
Honolulu, HI 96809
    (808) 548-7533
    Chief Engineer: Robert Chuck

Hawaii
Institute of Marine Biology
University of Hawaii
P.O. Box 1346
Kaneohe, HI 96744
    (808) 247-6631
    Director: William Coops

Hawaii
State Department of Health
Environmental Health
P.O. Box 3378
Honolulu, HI 96801
    (808) 548-4139
    Deputy Director: Dr. James Kumagai

Hawaii
State Department of Health
Environmental Protection Division
Pollution Technical Review Branch
P.O. Box 3378
Honolulu, HI 96801
    (808) 548-6478
    Environmental Engineer: William Wong

Hawaii
Water Resources Research Center
University of Hawaii
2540 Dole Street
Honolulu, HI 96822
    (808) 948-7847
    Director: L.S. Lau

Idaho
Department of Health and Welfare
Environmental Services Division
Water Quality Bureau
Statehouse
Boise, ID 83720
    (208) 384-2433
    Chief: Al Murray

Illinois
Environmental Protection Agency
Division of Public Water Supply
4500 South Sixth Street
Springfield, IL 62706
    (217) 785-0252
    Manager: Ira Markwood

Illinois
Environmental Protection Agency
Environmental Programs Division
2200 Churchill Road
Springfield, IL 62706
    (217) 782-3397
    Manager: Michael Mauzy

Indiana

Bureau of Engineering
Stream Pollution Control Board
1330 West Michigan Street
Indianapolis, IN 46206
(317) 633-0167
Director: Oral Hert

Indiana
Department of Natural Resources
Division of Water
608 State Office Building
Indianapolis, IN 46204
(317) 633-6344
Head: Robert Jackson

Indiana
State Board of Health
Division of Sanitary Engineering
1330 West Michigan Street
Indianapolis, IN 46206
(317) 633-0710
Interim Director: Chester Canham

Iowa
Department of Environmental Quality
Chemicals and Water Quality Division
Wallace Building
900 East Grand
Des Moines, IA 50319
(515) 265-8862
Director: Edward Brown

Kansas
Department of Health and Environment
Division of Environment
Bureau of Public Water Supply
Topeka, KS 66620
(913) 862-9360
Jack Burris

Kansas
Department of Health and Environment
Division of Environment
Bureau of Water Quality
Topeka, KS 66620
(913) 862-9360
Director: Eugene Jensen

Kentucky
Department for Natural Resources and
Environmental Protection
Division of Sanitary Engineering

Century Plaza
U.S. 127 South
Frankfort, KY 40601
(502) 564-3771
Director: Nick Johnson

Kentucky
Department for Natural Resources and
Environmental Protection
Division of Water Quality
Century Plaza
U.S. 127 South
Frankfort, KY 40601
(502) 564-3410
Director: Dr. Robert Blanz

Louisiana
Stream Control Commission
P.O. Drawer FC
University Station
Baton Rouge, LA 70893
(504) 342-6363
Executive Secretary: Robert Lafleur

Maine
Department of Environmental Protection
Bureau of Water Quality Control
State House
Augusta, ME 04333
(207) 289-2591
Director: Stephen Groves

Maine
Department of Human Services
Division of Environmental Health
State House
Augusta, ME 04330
(207) 289-0433
Director: Donald Hoxie

Maine Technology Society
1730 M Street, N.W.
Suite 412
Washington, DC 20036
(202) 659-3251
President: Dr. James Rickard

Maryland
Department of Natural Resources
Water Resources Administration
Tawes State Office Building
Annapolis, MD 21401

(301) 269-3776
Director: Lester Levine

Maryland
Environmental Health Administration
201 West Preston Street
Baltimore, MD 21201
  (301) 383-2740
  Environmental Health Administrator:
    Dr. Max Eisenberg

Maryland
State Department of Health and Mental
  Hygiene
Division of Water Supply
Water and Sewage Program
201 West Preston Street
Baltimore, MD 21201
  (301) 383-4249
  Chief: Raymond Anderson

Massachusetts
Department of Environmental Quality
  Engineering
Division of Water Supply
600 Washington Street
Boston, MA 02111
  (617) 727-2692
  Director: George Coogan

Massachusetts
Executive Office of Environmental
  Affairs
Department of Environmental Manage-
  ment
Division of Water Pollution Control
100 Cambridge Street
Boston, MA 02202
  (617) 727-3855
  Director: Thomas McMahon

Massachusetts
Executive Office of Environmental
  Affairs
Department of Environmental Manage-
  ment
Division of Water Resources
100 Cambridge Street
Boston, MA 02202
  (617) 727-3163
  Director: Charles Kennedy

Massachusetts
Executive Office of Environmental
  Affairs
Department of Metropolitan District
  Commission
Division of Water
20 Sommerset Street
Boston, MA 02108
  (617) 727-5114
  Director: James Matera

Michigan
Department of Natural Resources
Water Quality Division
Box 30028
Lansing, MI 48909
  (517) 373-1220
  Chief: Robert Courchaine

Michigan
Department of Public Health
Water Supply Division
Bureau of Environmental Health
3500 North Logan Street
Lansing, MI 48914
  (517) 373-1376
  Chief: William Kelly

Michigan
Water Resources Commission
Mason Building
Lansing, MI 48926
  (517) 373-2682
  Executive Secretary: Robert
    Courchaine

Minnesota
Department of Health
Water Supply and General Engineering
Division of Environmental Health
715 Delaware Street, S.W.
Minneapolis, MN 55440
  (612) 296-5330
  Chief: Gary England

Minnesota
Department of Natural Resources
Division of Water
658 Cedar Street
St. Paul, MN 55155
  (612) 296-4810
  Director: Larry Seymour

Minnesota
Pollution Control Agency
Division of Water Pollution Control
1935 West County Road B2
Roseville, MN 55113
  (612) 296-7315
  Acting Director: Lou Briemhurst

Minnesota
Water Resources Board
Room 206
555 Wabasha Street
St. Paul, MN 55102
  (612) 296-2840
  Chairman: Robert Moline

Mississippi
Air and Water Pollution Control
  Commission
P.O. Box 827
Jackson, MS 39205
  (601) 354-2550
  Chief: Charles Branch

Mississippi
Marine Conservation Commission
1201 East Bayview
Biloxi, MS 39530
  (601) 374-3205
  Director: Richard Leard

Mississippi
State Board of Health
Division of Water Supply
P.O. Box 1700
Jackson, MS 39205
  (601) 354-6616
  Director: James McDonald

Mississippi
State Board of Water Commissioners
416 North State Street
Jackson, MS 39201
  (601) 354-7236
  State Water Engineer: Jack Pepper

Missouri
Division of Environmental Quality
Water Quality Program
P.O. Box 1368
Jefferson City, MO 65101
  (314) 751-3241

Director: Richard Rankin

Missouri
Division of Environmental Quality
Water Supply Program
P.O. Box 1368
Jefferson City, MO 65101
  (314) 751-3241
  Director: Robert Miller

Montana
Department of Health and Environmental
  Sciences
Water Quality Bureau
Cogswell Building
Helena, MT 59601
  (406) 587-2406
  Chief: Donald Williams

Montana
Department of Health and Environmental
  Sciences
Water Quality Bureau
Potable Water Supply
Cogswell Building
Helena, MT 59601
  (406) 587-2406
  Chief: Arthur Clarkson

Montana
Department of Natural Resources and
  Conservation
Water Resources Division
32 South Ewing
Helena, MT 59601
  (406) 449-2872
  Administrator: Orrin Ferris

National Water Resources Association
955 L'Enfant Plaza North, S.W.
Suite 1202
Washington, DC 20024
  (202) 488-0610
  President: Hubert White

Nebraska
Department of Environmental Control
Water Pollution Control Division
301 Centennial Mall South
P.O. Box 94947
Lincoln, NE 68509
  (402) 471-2341

Chief: Bob Wall

Nebraska
Department of Health
Division of Environmental Engineering
Lincoln Building
10th and O Streets
Lincoln, NE 68509
  (402) 471-2674
  Director: Clifford Summers

Nebraska
Department of Water Resources
State House Station
Box 94676
Lincoln, NE 68509
  (402) 471-2363
  Director: John Neuberger

Nevada
Bureau of Consumer Health Protection
  Services
201 South Fall Street
Carson City, NV 89710
  (702) 885-4670
  Water Quality Officer: Wendell
    McCurry

Nevada
Department of Conservation and
  Natural Resources
Division of Water Planning
201 South Fall Street
Capitol Complex
Carson City, NV 89710
  (702) 885-4877
  Administrator: James Hawke

Nevada
Department of Conservation and
  Natural Resources
Division of Water Resources
Capitol Complex
201 South Fall Street
Carson City, NV 89710
  (702) 885-4380
  State Engineer: Roland Westergard

New England Interstate Water Pollution Control Commission
607 Boylston Street
Boston, MA 02116

  (617) 261-3758
  Executive Secretary: Alfred Peloquin

New Hampshire
Water Supply and Pollution Control
  Commission
105 Loudon Road
P.O. Box 95
Concord, NH 03301
  (603) 271-3503
  Executive Director: William Healy

New Jersey
Department of Environmental Protection
Bureau of Potable Water
1474 Prospect Street
P.O. Box 2809
Trenton, NJ 08625
  (609) 292-5550
  Acting Chief: Raymond Barg

New Jersey
Department of Environmental Conser-
  vation
Division of Marine Services
P.O. Box 1889
Trenton, NJ 08625
  (609) 292-2795
  Director: Donald Graham

New Jersey
Department of Environmental Protection
Division of Water Resources
1474 Prospect Street
P.O. Box 2809
Trenton, NJ 08625
  (609) 292-1637
  Deputy Director: Jeff Zelikson

New Mexico
Environmental Improvement Agency
Water Quality Division
P.O. Box 2348
Sante Fe, NM 87503
  (505) 827-5271
  Chief: Joseph Pierce

New Mexico
Environmental Improvement Agency
Water Supply Section
P.O. Box 2348
Sante Fe, NM 87503

(505) 827-5271
Chief: Francisco Garcia

New York
Bureau of Public Water Supply
Division of Sanitary Engineering
Empire State Plaza
Tower Building
Albany, NY 12237
  (518) 474-5577
    Associate Director: Samuel
      Syrotynski

New York
Department of Environmental Conservation
Division of Pure Waters
50 Wolf Road
Albany, NY 12233
  (518) 457-6674
    Director: Eugene Seebald

New York
Marine Sciences Research Center
State University of New York
Stony Brook, NY 11794
  (516) 246-7710
    Director: Dr. J.R. Schubel

New York Ocean Science Laboratory
Public Information Office
Edgemere Road
Montauk, NY 11954
  (516) 668-5800
    Byron Porterfield

North Carolina
Department of Human Resources
Division of Health Services
Sanitary Engineering Section
Bath Building
P.O. Box 2091
Raleigh, NC 27602
  (919) 733-2870
    Head: Charles Rundgren

North Carolina
Department of Natural Resources and
  Community Development
Division of Environmental Management
Water Quality Division
P.O. Box 27687

Raleigh, NC 27611
  (919) 733-5083
    Chief: L. Page Benton, Jr.

North Dakota
Department of Health
Division of Water Supply and Pollution
  Control
Missouri Office Building
Bismark, ND 58505
  (701) 224-2386
    Director: Norman Peterson

Oceanic Society
Fort Mason Building 240
San Francisco, CA 94123
  (415) 441-1104
    President

Ohio
Environmental Protection Agency
Division of Water Quality Standards
P.O. Box 1049
Columbus, OH 43216
  (614) 466-6686
    Chief: Ernie Rotering

Ohio
Environmental Protection Agency
Office of Public Water Supply
P.O. Box 1049
Columbus, OH 43216
  (614) 466-8307
    Chief: James Kneale

Ohio
Environmental Protection Agency
Water Quality Planning and Assessment
Box 1049
Columbus, OH 43216
  (614) 466-6533
    Pete Digman

Oklahoma
Department of Health
Water Quality Service
N.E. 10th and Stonewall
Oklahoma City, OK 73117
  (405) 271-6315
    Chief: Charles Newton

Oregon
Department of Environmental Quality
522 S.W. 5th Avenue
P.O. Box 1760
Portland, OR 97207
    (503) 229-6403
    Water Quality Coordinator: Glen
      Carter

Oregon
State Health Division
Office of Protective Health Services
1400 S.W. 5th Avenue
Portland, OR 97201
    (503) 229-5681
    Chief: Douglas Pike

Pennsylvania
Department of Environmental Resources
Bureau of Water Quality Management
P.O. Box 17120
Harrisburg, PA 17120
    (717) 787-2666
    Acting Director: Lou Berchini

Pennsylvania
Department of Environmental Resources
Division of Water Supply and Sewage
P.O. Box 3780
Harrisburg, PA 17120
    (717) 787-3481
    Director: Therold Krammes

Rhode Island
Department of Environmental Manage-
    ment
Division of Water Resources
Room 209
Cannon Building
75 Davis Street
Providence, RI 02908
    (401) 277-2234
    Chief: James Fester

Rhode Island
Department of Health
Health Building
Room 209
75 Davis Street
Providence, RI 02908
    (401) 277-2234
    Sanitary Engineer: John Hagopian

River Conservation Fund
317 Pennsylvania Avenue, S.E.
Washington, DC 20003
    (202) 547-6900
    President: Howard Brown

South Carolina
Department of Health and Environmental
    Control
Water Supply Division
2600 Bull Street
Columbia, SC 29201
    (803) 758-5544
    Director: Lewis Shaw

South Carolina
Environmental Quality Control
Bureau of Wastewater and Stream
    Quality
2600 Bull Street
Columbia, SC 29201
    (803) 758-3877
    Chief: Charles Jeter

South Dakota
Department of Environmental Protec-
    tion
Water Quality Control
Water Hygiene Program
Pierre, SD 57501
    (605) 773-3351
    Chief: Mark Steichen

South Dakota
Department of Natural Resource Devel-
    opment
Division of Water Rights
Joe Foss Office Building
Pierre, SD 57501
    (605) 773-3151
    Director: John Hatch

Tennessee
Department of Public Health
Division of Water Quality Control
621 Cordell Hill Building
Nashville, TN 37219
    (615) 741-2275
    Director: Elmo Lunn

Tennessee
Department of Public Health

Drinking Water Quality Control
320 Capitol Hill Building
Nashville, TN 37219
  (615) 741-2281
  Chief: David Droughon

Texas
Department of Health
Water Hygiene Division
1100 West 49th Street
Austin, TX 78756
  (512) 458-7111
  Director: Charles Foster

Texas
Department of Water Resources
P.O. Box 13087
Capitol Station
Austin, TX 78711
  (512) 475-3187
  Executive Director: Harvey Davis

Texas
Department of Water Resources
Water Commission
P.O. Box 13087
Capitol Station
Austin, TX 78711
  (512) 475-4514
  Chief Hearing Examiner: Lee Mathews

Texas
Guadalupe-Blanco River Authority
Water Quality Services
P.O. Box 271
Seguin, TX 78155
  (512) 379-5822
  Supervisor: James Arnst

U.S.
Committee on Large Dams
Environmental Effects Committee
345 East 47th Street
New York, NY 10017
  (212) 644-7848
  Chairman: Wallace Chadwick

U.S.
Department of Agriculture
Department of Science and Education
  Administration
Southwest Watershed Research Center

442 East Seventh Street
Tucson, AZ 85705
  (602) 792-6381
  Kenneth Renard

U.S.
Department of Agriculture
Science and Education Administration
Water Quality Management Laboratory
Wilson Street
Route 2
Box 322A
Durant, OK 74701
  (405) 924-5066
  Ronald Menzel

U.S.
Department of Agriculture
Soil Conservation Service
Water Resources
P.O. Box 2890
Washington, DC 20013
  (202) 447-4527
  Assistant Administrator: Joseph
    Haas

U.S.
Department of Commerce
NOAA
National Ocean Survey
Atlantic Marine Center
439 West York Street
Norfolk, VA 23510
  (804) 441-6439

U.S.
Department of Commerce
NOAA
National Ocean Survey
Pacific Marine Center
1801 Fairview Avenue, East
Seattle, WA 98102
  (206) 442-7657

U.S.
Department of the Interior
Bureau of Land Management
Marine Environmental Assessment
  Branch
18th and C Streets, N.W.
Washington, DC 20420
  (202) 343-6264

Chief

Thomas Jorlinq

U.S.
Department of the Interior
Fish and Wildlife Service
National Stream Alteration Team
Route 1
Columbia, MO 65201
   (314) 442-2271

U.S.
Department of the Interior
Geological Survey
National Water Data Exchange
12201 Sunrise Valley Drive
Reston, VA 22092
   (703) 860-6031

U.S.
Department of the Interior
Office of Environmental Project
   Review
Water Resources
Office of the Secretary
Washington, DC 20240
   (202) 343-5464
   Staff: Vincent Sullivan

U.S.
Department of the Interior
Office of Water Research and Tech-
   nology
Washington, DC 20240
   (202) 343-4608
   Director: Gary Cobb

U.S.
Environmental Protection Agency
EIS Preparation Section
Water Division
Region V
230 Dearborn Street
Chicago, IL 60604
   (312) 353-2000
   Section Chief: Eugene Wojcik

U.S.
Environmental Protection Agency
Water and Wastewater Management
401 M Street, S.W.
Washington, DC 20460
   (202) 755-2673

U.S.
Environmental Protection Agency
Water Planning and Standards
401 M Street, S.W.
Washington, DC 20460
   (202) 755-2673
   Deputy Assistant Administrator:
   Swep Davis

U.S.
Nuclear Regulatory Commission
Division of Site Safety and Environ-
   mental Analysis
Hydrology-Meteorology Branch
Washington, DC 20555
   (301) 492-8003
   Acting Chief: William Bivins

U.S.
Water Conservation Laboratory
4331 East Broadway
Phoenix, AZ 85040
   (602) 261-4356
   Director: Dr. Herman Bouwer

U.S.
Water Resources Council
2120 L Street, N.W.
Washington, DC 20037
   (202) 254-6303
   Policy Officer: Dave Shepard

U.S.
Water Resources Council
Public Information Office
2120 L Street, N.W.
Suite 800
Washington, DC 20037
   (202) 254-6303
   Officer: Wanda Phelan

Utah
Bureau of Water Quality
150 W.N. Temple
Room 426
Salt Lake City, UT 84110
   (801) 533-6146
   Director: Calvin Sudweeks

Utah
Environmental Coordinating Committee
Division of Water Resources
124 State Capitol
Salt Lake City, UT   84111
    (801) 533-5794
    Barry Saunders

Utah
State Department of Natural Resources
Division of Water Resources
Empire Building
Suite 300
231 East Fourth South
Salt Lake City, UT   84111
    (801) 533-5401
    Director: Daniel Lawrence

Utah
State Division of Health
Bureau of Water Quality
44 Medical Drive
Salt Lake City, UT   84113
    (801) 533-6111
    Director

Vermont
Agency of Environmental Conservation
Water Resources Department
Montpelier, VT   05602
    (802) 828-3361
    Reginald LaRosa

Vermont
Agency of Environmental Conservation
Water Quality Section
State Office Building
Montpelier, VT   05602
    (802) 826-2761
    Director: David Clough

Vermont
Department of Health
Division of Environmental Health
60 Main Street
Burlington, VT   05401
    (802) 862-5701
    Chief: Kenneth Stone

Virginia
Department of Health

Office of Health Protection and Environmental Management
Division of Water Management
109 Govenor Street
Richmond, VA   23219
    (804) 786-6277
    Director: Eric Bartsch

Virginia
Institute of Marine Science
Gloucester Point, VA   23062
    (804) 642-2111
    Director: William Hargis, Jr.

Virginia
Marine Resources Commission
P.O. Box 756
2401 West Avenue
Newport News, VA   23607
    (804) 245-2811
    Commissioner: James Douglas, Jr.

Virginia
State Water Control Board
2111 North Hamilton Street
Richmond, VA   23230
    (804) 257-0056
    Executive Secretary: Robert Davis

Virginia
State Water Control Board
Bureau of Surveillance and Field
  Studies
2111 North Hamilton Street
Richmond, VA   23230
    (804) 257-0389

Virgin Islands
Department of Conservation and Cul-
  tural Affairs
Division of Natural Resource Manage-
  ment
P.O. Box 4340
St. Thomas, VI   00801
    (809) 774-6420
    Director: Pedrito Francois

Washington
Department of Ecology
Office of Water Programs
Olympia, WA   98504

(206) 753-3893
Assistant Director: John Spencer

Washington
Department of Ecology
Water Quality Management Division
Olympia, WA 98504
  (206) 753-2800
  Division Supervisor: Glen Fiedler

Washington
Department of Ecology
Water Resources Management Division
Olympia, WA 98504
  (206) 753-2800
  Division Supervisor: Gene Wallace

Water Management Association of Ohio
445 King Avenue
Columbus, OH 43201
  (614) 424-6106
  President: Bennett Coy

Water Pollution Control Federation
2626 Pennsylvania Avenue, N.W.
Washington, DC 20037
  (202) 337-2500
  President: Martin Lang

Water Resources Association of the
  Delaware River Basin
Box 867
Davis Road
Valley Forge, PA 19481
  (215) 783-0634
  President: Robert Copeland

West Virginia
Department of Natural Resources
Division of Water Resources
1201 Greenbrier Street
Charleston, WV 25311
  (304) 348-2107
  Chief: David Robinson

West Virginia
State Department of Health
Environmental Health Services
State Office Building
1800 East Washington Street
Charleston, WV 25305
  (304) 348-2981

Director: Robert McCall

Wisconsin
Department of Natural Resources
Division of Environmental Studies
Public Water Supply Section
P.O. Box 450
Madison, WI 53701
  (608) 266-2299
  Chief: Robert Baumeister

Wyoming
Department of Environmental Quality
Water Quality Division
Hathaway Building
Cheyenne, WY 82002
  (307) 328-9781
  Administrator: William Garland

CULTURAL DIRECTORIES

8. ARCHAEOLOGY/HISTORY-RELATED AGENCIES AND ORGANIZATIONS

Alabama
Department of Archives and History
Historic Preservation
Archives and History Building
Montgomery, AL 36104
  (205) 832-6510
  Director: Milo Howard, Jr.

Alaska
Division of Parks
Office of History and Archaeology
619 Warehouse Avenue
Suite 210
Anchorage, AK 99501
  (907) 274-4676
    State Historic Preservation
      Officer: William Hanable

American Association for State and
  Local History
1400 Eighth Avenue, South
Nashville, TN 37203
  (615) 242-5583

American Historical Association
400 A Street, S.E.
Washington, DC 20003
  (202) 544-2422

American Scenic and Historic Pres-
  ervation Society
35 West 53rd Street
Penthouse
New York, NY 10019
  President: John Pell

Archaeology Institute of America
53 Park Place
New York, NY 10007
  (212) 732-6677

Architectural Heritage
Old City Hall
45 School Street
Boston, MA 02108
  (617) 523-8678

Arizona
State Parks
Historic Preservation
1688 West Adams
Phoenix, AZ 85007
  (602) 255-4174
    State Historic Preservation
      Officer: James Ayres

Arkansas
Historic Preservation Program
Suite 500
Continental Building
Markham and Main Streets
Little Rock, AR 72201
  (501) 371-2763
    State Historic Preservation
      Officer: Joan Williams Baldridge

California
Department of Parks and Recreation
Office of Historic Preservation
P.O. Box 2390
1220 K Street

Sacramento, CA 95811
  (916) 322-8596
  Director: Knox Mellon

Colorado Historical Society
Historic Preservation
1300 Broadway
Denver, CO 80203
  (303) 839-3397
  Chairman: Arthur Townsend

Connecticut Historical Commission
Historic Preservation
59 South Prospect Street
Hartford, CT 06106
  (203) 566-3005
  Director: John Shannahan

Delaware
Division of Historical and Cultural
  Affairs
Historic Preservation
Hall of Records
Dover, DE 19901
  (303) 736-5314
  Director: Lawrence Henry

District of Columbia
Department of Housing and Community
  Development
Historic Preservation
1325 G Street, N.W.
Washington, DC 20005
  (202) 724-2120
  Director: Robert Moore

Florida
Department of State
Division of Archives, History and
  Records Management
Historic Preservation
401 East Gaines Street
Tallahassee, FL 32304
  (904) 488-1480
    State Historic Preservation
      Officer: Ross Morrell

Georgia
Department of Natural Resources
Historic Preservation Section
Room 701
270 Washington Street, S.W.

Atlanta, GA 30334
  (404) 656-2840
  Chief: Elizabeth Lyon

Guam
Government of Guam
Department of Parks and Recreation
Historic Preservation
P.O. Box 2950
Agana, Guam 96910
  (Overseas Operator) 477-9620
  Director: Felix Crisostomo

Hawaii
Department of Land and Natural Re-
  sources
P.O. Box 621
Honolulu, HI 96809
  (808) 548-6550
    State Historic Preservation
      Officer: Susumu Ono

Idaho Historical Society
Historic Preservation
610 North Julia Davis Drive
Boise, ID 83702
  (208) 334-3356
    State Historic Preservation
      Officer: Merle Wells

Illinois
Department of Conservation
Historic Preservation
602 Stratton Building
Springfield, IL 62706
  (217) 782-6302
  Director: David Kenney

Indiana
Division of Historic Preservation
202 North Alabama Street
Indianapolis, IN 46204
  (317) 232-1646
  Deputy: Richard Gantz

Iowa
State Historical Department
Division of Historic Preservation
26 East Market Street
Iowa City, IA 52240
  (319) 353-4186
  Director: Adrian Anderson

Kansas State Historical Society
Historic Preservation
120 West Tenth
Topeka, KS 66612
  (913) 296-3251
  Executive Director: Joseph Snell

Kentucky Heritage Commission
Historic Preservation
Old YMCA Building
104 Bridge Street
Frankfort, KY 40601
  (502) 564-6683
  Director: Eldred Melton

Louisiana
Office of Program Development
Historic Preservation
P.O. Box 44247
Baton Rouge, LA 70804
  (504) 925-3880
  Assistant Secretary: Bernard
    Carrier

Maine Historic Preservation
  Commission
242 State Street
Augusta, ME 04333
  (207) 289-2133
  Director: Earle Shettleworth, Jr.

Maryland
Historic Preservation
John Shaw House
21 State Circle
Annapolis, MD 21401
  (301) 269-2440
    State Historic Preservation
      Officer: Rodney Little

Massachusetts Historical Commission
Historic Preservation
294 Washington Street
Boston, MA 02108
  (617) 727-8470
    State Historic Preservation
      Officer: Patricia Weslowski

Michigan
Department of State
Michigan History Division
Historic Preservation

208 North Capitol
Lansing, MI 48918
  (517) 373-6362
  Director: Martha Bigelow

Minnesota Historical Society
Historic Preservation
690 Cedar Street
St. Paul, MN 55101
  (612) 296-2747
  Director: Russell Fridley

Mississippi
Department of Archives and History
Historic Preservation
P.O. Box 571
Jackson, MS 39205
  (601) 354-6218
  Director: Elbert Hilliard

Missouri
Historic Preservation
1001 Southwest Boulevard
P.O. Box 176
Jefferson City, MO 65102
  (314) 751-4096
  Deputy: Orval Henderson, Jr.

Montana Historical Society
Historic Preservation
225 North Roberts Street
Veterans' Memorial Building
Helena, MT 59601
  (406) 449-2694
  Director: Robert Archibald

National Conference of State Historic
  Preservation Officers
1522 K Street, N.W.
Suite 500
Washington, DC 20005
  (202) 254-3974

National Trust for Historic Preservation
748 Jackson Place, N.W.
Washington, DC 20006
  (202) 638-5200

Natural History Society of Maryland
2643 North Charles Street
Baltimore, MD 21218

(301) 235-6116
President: Elra Palmer

Natural History Society of Puerto
  Rico
Box 1393
Hato Rey, PR 00919
  (809) 789-9371
  President: Jose Castillo

Nebraska State Historical Society
Historic Preservation
1500 R Street
Lincoln, NE 68508
  (402) 471-3270
  Director: Marvin Kivett

Nevada
Department of Conservation and
  Natural Resources
Division of Historic Preservation
  and Archaeology
Nye Building
201 South Fall Street
Carson City, NV 89710
  (702) 885-5138
  State Historic Preservation
    Officer: Mimi Rodden

New Hampshire
Department of Resources and
  Economic Development
Historic Preservation
P.O. Box 856
Concord, NH 03301
  (603) 271-2411
  Deputy: Linda Wilson

New Jersey
Department of Environmental Conser-
  vation
Division of Parks and Forestry
Historic Sites Section
P.O. Box 1420
Trenton, NJ 08625
  (609) 292-2023
  Supervisor: Judith Blood

New Mexico
State Planning Office
Historic Preservation
P.O. Box 1629

505 Don Gaspar
Santa Fe, NM 87503
  (505) 827-2108
  State Historic Preservation
    Officer: Thomas Merlan

New York
State Office of Parks and Recreation
Historic Preservation
Empire State Plaza
Albany, NY 12238
  (518) 474-0456
  Deputy Commissioner for Historic
    Preservation: Frederick Rath, Jr.

North Carolina
Department of Cultural Resources
Division of Archives and History
Historic Preservation
109 East Jones Street
Raleigh, NC 27611
  (919) 733-7305
  Director: Larry Tise

North Dakota
State Historical Society of North
  Dakota
Historical Preservation
Liberty Memorial Building
Bismark, ND 58501
  (701) 224-2667
  Superintendent: James Sperry

Ohio Historical Society
Historic Preservation
Interstate 71 at 17th Avenue
Columbus, OH 43211
  (614) 466-1500
  State Historic Preservation
    Officer: David Brook

Oklahoma
Historic Preservation
Historical Building
2100 Lincoln, North
Oklahoma City, OK 73105
  (405) 521-2491
  State Historic Preservation
    Officer: Glenn Jordan

Oregon
State Parks

Historic Preservation
525 Trade Street, S.E.
Salem, OR 97310
  (502) 378-5019
  Deputy: David Powers

Pennsylvania Historical and Museum
  Commission
Historic Preservation
P.O. Box 1026
Harrisburg, PA 17120
  (717) 787-4363
  State Historic Preservation
    Officer: Ed Weintraub

Puerto Rico
Office of Cultural Affairs
Historic Preservation
La Fortaleza
San Juan, PR 00905
  (809) 724-2100
  State Historic Preservation
    Officer: Rafael Rivera Garcia

Rhode Island Historic Preservation
  Commission
Old State House
150 Benefit Street
Providence, RI 02903
  (401) 277-2678
  Deputy: Eric Hertfelder

Society for Architectural Historians
1700 Walnut Street
Room 716
Philadelphia, PA 19103
  (215) 735-0224
  Executive Secretary

South Carolina
State Archives Department
Historic Preservation
1430 Senate Street
Columbia, SC 29211
  (803) 758-5816
  Deputy: Christie Fant

South Dakota Historic Preservation
  Center
University of South Dakota
Alumni House
Vermillion, SD 57069

(605) 677-5314
State Historic Preservation
  Officer: John Little

Tennessee Historical Commission
Historic Preservation
4721 Trousdale Drive
Nashville, TN 37220
  (615) 741-2371
  Deputy: Paul Cross

Texas Historical Commission
Historic Preservation
P.O. Box 12276
Capitol Station
Austin, TX 78711
  (512) 475-3092
  Executive Director: Truett Latimer

U.S.
Advisory Council on Historic Preser-
  vation
Office of Review and Compliance
1522 K Street, N.W.
Suite 430
Washington, DC 20005
  (202) 254-3380
  Assistant Director: Myra Harrison

U.S.
Department of the Interior
Heritage Conservation and Recreation
  Service
C Street Between 18th and 19th
  Streets, N.W.
Washington, DC 20240
  (202) 343-1100
  Director: Chris Delaporte

U.S.
Department of the Interior
Heritage Conservation and Recreation
  Service
Division of Environmental and Com-
  pliance Review
440 G Street, N.W.
Washington, DC 20243
  (202) 343-4275

U.S.
Department of the Interior

Heritage Conservation and Recreation
  Service
Programming Office
19th and C Streets, N.W.
Washington, DC 20240
  (202) 343-5711
  Program Liaison Specialist

U.S.
Department of the Interior
Heritage Conservation and Recreation
  Service
Alaska Regional Office
540 West Fifth Street
Anchorage, AK 99501
  (907) 265-5345

U.S.
Department of the Interior
Heritage Conservation and Recreation
  Service
Lake Central Regional Office
3853 Research Park Drive
Ann Arbor, MI 48104
  (313) 769-3211

U.S.
Department of the Interior
Heritage Conservation and Recreation
  Service
Mid-Continent Regional Office
Denver Federal Center
Building 41
P.O. Box 25387
Denver, CO 80225
  (303) 234-2634

U.S.
Department of the Interior
Heritage Conservation and Recreation
  Service
Northeast Regional Office
600 Arch Street
Philadelphia, PA 19106
  (215) 597-7990

U.S.
Department of the Interior
Heritage Conservation and Recreation
  Service
Northwest Regional Office
450 Golden Gate Avenue

San Francisco, CA 94102
  (415) 556-0182

U.S.
Department of the Interior
Heritage Conservation and Recreation
  Service
South Central Regional Office
5000 Marble Avenue, N.E.
Alburquerque, NM 87110
  (505) 766-3515

U.S.
Department of the Interior
Heritage Conservation and Recreation
  Service
Southeast Regional Office
148 International Boulevard
Atlanta, GA 30303
  (404) 526-4405

U.S.
Department of the Interior
National Park Service
Archaeology and Historic Preservation
1100 L Street, N.W.
Washington, DC 20240
  (202) 523-5275
  Director

U.S.
Department of the Interior
National Park Service
Cultural Resources Management
  Division
18th and C Streets, N.W.
Washington, DC 20240
  (202) 343-7550
  Director

U.S.
National Trust for Historic Preser-
  vation
Preservation Services Office
740-748 Jackson Place, N.W.
Washington, DC 20006
  (202) 638-5200
  Vice President: Russell Keune

Utah
Environmental Coordinating Committee
Division of State History

124 State Capitol
Salt Lake City, UT 84111
  (801) 533-5794
  Jim Dykman

Utah State Historical Society
Historic Preservation
307 West 200 South
Suite 1000
Salt Lake City, UT 84101
  (801) 533-5755
  Director: Melvin Smith

Vermont
Agency of Development and Community
  Affairs
Historic Preservation
Pavilion Office Building
Montpelier, VT 05602
  (802) 828-3226
  Deputy: William Pinney

Virginia Historic Landmarks Commission
Historic Preservation
221 Govenor Street
Richmond, VA 23219
  (804) 786-3143
  Executive Director: Tucker Hill

Washington
Office of Archaeology and Historic
  Preservation
111 West 21st Street
Olympia, WA 98504
  (206) 753-6780
  Director: Louis Guzzo

West Virginia
Department of Culture and History
Historic Preservation
Capitol Complex
Charleston, WV 25304
  (304) 348-0244
  State Historic Preservation
    Officer: Clarence Moran

Wisconsin State Historical Society
Historic Preservation
816 State Street
Madison, WI 53706
  (608) 262-9504
  Deputy: Jeffrey Dean

Wyoming Recreation Commission
Historic Preservation
604 East 25th Street
Box 309
Cheyenne, WY 82001
  (307) 777-7695
  Deputy: John Carlson

9. ENERGY/UTILITY-RELATED AGENCIES AND ORGANIZATIONS

Alabama
Department of Public Health
Division of Solid Waste and Vector
  Control
State Office Building
Montgomery, AL 36104
  (205) 832-6728
  Director: Alfred Chipley

Alaska
Department of Environmental Con-
  servation
Solid Waste Management
Pouch O
Juneau, AK 99811
  (907) 465-2672
  Dick Williams

American Nuclear Society
Environmental Sciences Division
555 North Kensington Avenue
La Grange Park, IL 60525
  (312) 352-6611
  Division Chairman

American Nuclear Society
Public Information Office
555 North Kensington Avenue
La Grange Park, IL 60525
  (312) 352-6611
  Manager: Ed Ronne

American Petroleum Institute
Central Abstracting and Indexing
  Service

275 Madison Avenue
New York, NY 10016
  (212) 685-6254
  Assistant Manager: I. Zarember

Arizona
Bureau of Sanitation
1740 West Adams Street
Phoenix, AZ 85007
  (602) 255-1156
  Chief: John Beck

Arkansas
Department of Pollution Control and
  Ecology
Solid Waste Division
8001 National Drive
Little Rock, AR 72209
  (501) 371-1701
  Chief: Buddy Parr

California
Energy Resources Conservation and
  Development Commission
1111 Howe Avenue
Room 331-B
Sacramento, CA 95825
  (916) 920-6405
  Justin Tierney

California
Public Utilities Commission
350 McAllister Street
San Francisco, CA 94102

(415) 557-0442
Donald Steger

California
Solid Waste Management Board
825 K Street
Suite 300
Sacramento, CA 95814
  (916) 322-2658
  Bob Sleppe

Colorado
Department of Health
Solid Waste Management Project
4210 East 11th Avenue
Denver, CO 80220
  (303) 388-6111
  Orville Stoddard

Colorado
Department of Health
Water Quality Division
4210 East 11th Avenue
Denver, CO 80220
  (303) 320-8333
  Industrial Waste Consultant:
    Robert Shukle

Connecticut
Department of Environmental Conser-
  vation
Solid Waste Management Programs
122 Washington Street
Hartford, CT 06115
  (203) 566-3672
  Director: Charles Kurker

Delaware
Solid Waste Authority
P.O. Box 981
Dover, DE 19901
  (302) 678-5361
  General Manager: N.C. Vasuki

District of Columbia
Solid Waste Administration
415 12th Street, N.W.
Room 303
Washington, DC 20004
  (202) 629-4581
  Deputy Administrator: William
    Johnson

Electric Power Research Institute
P.O. Box 10412
Palo Alto, CA 94303
  (415) 855-2000
  Vice Chairman: Chauncey Starr

Environmental Law Institute
Energy Program
1346 Connecticut Avenue, N.W.
Suite 600
Washington, DC 20036
  (202) 452-9600
  Director: Curtis Seltzer

Florida
Department of Environmental Regulation
Industrial Waste Section
Division of Environmental Permitting
2562 Executive Center Circle, East
Tallahassee, FL 32301
  (904) 488-3391
  Dr. James Brindell

Florida
Department of Environmental Regulation
Solid Waste Management Program
2562 Executive Center Circle, East
Tallahassee, FL 32301
  (904) 488-0300
  Administrator: William Buzick

Georgia
Department of Natural Resources
Environmental Protection Division
Solid Waste Management Service
270 Washington Street, S.W.
Room 822
Atlanta, GA 30334
  (404) 656-2833
  Director: Moses McCall, III

Idaho
Department of Health and Welfare
Solid Waste Management Section
Boise, ID 83720
  (208) 384-2390
  Director

Illinois
Environmental Protection Agency
Division of Enforcement Services
2200 Churchill Road

Springfield, IL 62706
  (217) 782-5544
  Senior Technical Advisor: Barbara
    Sidler

Illinois
Environmental Protection Agency
Division of Land-Noise Pollution
  Control
2200 Churchill Road
Springfield, IL 62706
  (217) 782-6760
  Manager: John Moore

Indiana
Department of Natural Resources
Division of Oil and Gas
608 State Office Building
Indianapolis, IN 46204
  (317) 633-6344
  Head: Homer Brown

Indiana
State Board of Health
Division of Sanitary Engineering
Solid Waste Section
1330 West Michigan Street
Indianapolis, IN 46206
  (317) 633-0176
  Acting Chief: David Lamm

Indiana
State Board of Health
Division of Water Pollution Control
1330 West Michigan Street
Indianapolis, IN 46206
  (317) 633-0700
  Director: Earl Bohner

Interstate Sanitation Commission
10 Columbus Circle
New York, NY 10019
  (212) 582-0380
  Director: Thomas Glenn, Jr.

Kansas
Department of Health and Environment
Division of the Environment
Bureau of Environmental Sanitation
Topeka, KS 66620
  (913) 862-9360
  Director: J. Howard Duncan

Kentucky
Department for Natural Resources and
  Environmental Protection
Division of Solid Waste
Pine Hill Plaza
U.S. 60
Frankfort, KY 40601
  (502) 564-6716
  John Smither

Long Island Lighting Company
Environmental Engineering Department
175 East Old Country Road
Hicksville, NY 11801
  (516) 733-4137
  Raymond Driscoll

Louisiana
Health and Human Resources Admin-
  istration
Solid Waste and Vector Control Unit
P.O. Box 60630
New Orleans, LA 70160
  (504) 568-5100
  Administrator: Roy Hayes, Jr.

Maine
Department of Environmental Protec-
  tion
Division of Industrial Services
Hospital Road
Augusta, ME 04333
  (207) 289-2591
  Chief: Leonard Rost

Maine
Department of Environmental Protec-
  tion
Division of Solid Waste Management
State House
Augusta, ME 04333
  (207) 289-2111
  Chief: Ron Howes

Maryland
Department of Health and Mental
  Hygiene
Division of Solid Waste
P.O. Box 13387
Baltimore, MD 21203
  (301) 383-2770
  Chief: Arthur Caple

Maryland
Department of Natural Resources
Energy and Coastal Zone Management
  Administration
Tawes State Office Building
Annapolis, MD 21401
  (301) 269-2788
  Director: Lee Zeni

Massachusetts
Executive Office of Environmental
  Affairs
Department of Environmental Manage-
  ment
Division of Solid Waste Disposal
100 Cambridge Street
Boston, MA 02202
  (617) 727-3163
  Director: William Gaughan

Massachusetts
Executive Office of Environmental
  Affairs
Department of Environmental Quality
  Engineering
Division of Water Pollution Control
100 Cambridge Street
Boston, MA 02202
  (617) 727-3855
  Glen Gilmore

Massachusetts
Executive Office of Environmental
  Affairs
Department of Metropolitan District
  Commission
Division of Sewage
20 Sommerset Street
Boston, MA 02108
  (617) 727-5114
  Director: Walter Paton

Metropolitan Sanitary District of
  Greater Chicago
Management Control Office
100 East Erie
Fifth Floor
Chicago, IL 60611
  (312) 751-5653
  Officer: William Coyne

Michigan
Department of Natural Resources
Resource Recovery Division
905c Southland
Lansing, MI 48926
  (517) 322-1315
  Chief: Fred Kellow

Minnesota
Pollution Control Agency
Division of Solid Wastes
1935 West County Road B2
Roseville, MN 55113
  (612) 296-7316
  Director: Robert Silvagni

Mississippi
Air and Water Pollution Control
  Commission
Industrial Waste Water Section
P.O. Box 1700
Jackson, MS 39205
  (601) 354-2550
  Jerry Banks

Mississippi
State Board of Health
Division of Sanitary Engineering
Solid Waste Planning Program
P.O. Box 1700
Jackson, MS 39201
  (601) 354-6616
  Director: Jack McMillian

Missouri
Division of Environmental Quality
Solid Waste Program
P.O. Box 1368
Jefferson City, MO 65101
  (314) 751-3241
  Director: Robert Robinson

Montana
Department of Health and Environmen-
  tal Sciences
Solid Waste Management Bureau
Cogswell Building
Helena, MT 59601
  (406) 449-2821
  Chief: Duane Robertson

Montana
Department of Natural Resource and
   Conservation
Energy Planning Division
32 South Ewing
Helena, MT 59601
   (406) 449-3780
   Administrator: Robert Anderson

National Petroleum Council
Information Office
1625 K Street, N.W.
Washington, DC 20006
   (202) 393-6100
   Director: Joan Walsh

National Solid Waste Management
   Association
1120 Connecticut Avenue, N.W.
Suite 930
Washington, DC 20036
   (202) 659-4613
   Technical Director

Nebraska
Department of Environmental Control
Division of Solid Waste Management
P.O. Box 94877
State House Station
Lincoln, NE 68509
   (402) 471-2186
   Chief: Bill Sheil

Nebraska
Department of Environmental Control
Permits and Enforcement Section
P.O. Box 94877
State House Station
Lincoln, NE 68509
   (402) 471-2186
   Jay Ringenberg

Nevada
Department of Conservation and
   Natural Resources
Solid Waste Management
201 South Fall Street
Carson City, NV 89710
   (702) 885-4670
   Program Director: Laverne Rosse

New Hampshire
Department of Health and Welfare
Division of Public Health
State Laboratory Building
Hazen Drive
Concord, NH 03301
   (603) 271-2605
   Thomas Sweeney

New Hampshire
Water Supply and Pollution Control
   Commission
105 Loudon Road
P.O. Box 95
Concord, NH 03301
   (603) 271-3503
   Senior Sanitary Engineer: Russell
   Nylander

New Jersey
Department of Environmental Protec-
   tion
Solid Waste Administration
P.O. Box 1390
Trenton, NJ 08625
   (609) 292-2697
   Director: Beatrice Tylutki

New Jersey
Department of Environmental Protec-
   tion
Water Planning and Management Element
1474 Prospect Street
P.O. Box 2809
Trenton, NJ 08625
   (609) 292-8427
   Assistant Director: Douglas Clark

New Mexico
Environmental Improvement Agency
Water Pollution Control
Surveillance and Standards Section
P.O. Box 2348
Santa Fe, NM 87503
   (505) 827-5271
   Program Manager: Michael Snavely

New York
Department of Environmental Conser-
   vation
Bureau of Industrial Programs

50 Wolf Road
Albany, NY 12233
   (518) 457-3967
   Director: Salvatore Pagano

New York
Department of Environmental Conser-
   vation
Division of Solid Waste Management
50 Wolf Road
Albany, NY 12233
   (518) 457-6603
   Director: William Bentley

North Carolina
Department of Human Resources
Division of Health Services
Solid Waste and Vector Control Pro-
   gram
P.O. Box 2091
Raleigh, NC 27602
   (919) 733-2178
   Head: Jerry Perkins

North Carolina
Department of Natural Resources and
   Community Development
Division of Environmental Management
Field Services Section
P.O. Box 27687
Raleigh, NC 27611
   (919) 733-4740
   Supervisor: Robert Carter

North Dakota
Department of Health
Division of Water Supply and Pollu-
   tion Control
Missouri Office Building
1200 Missouri Avenue
Bismark, ND 58505
   (701) 224-2375
   Director: Gerald Knudsen

Ohio
Environmental Protection Agency
Division of Industrial Waste Water
P.O. Box 1049
Columbus, OH 43216
   (614) 466-2390
   Assistant Chief: Andrew Turner

Ohio
Environmental Protection Agency
Power Siting Commission
Box 1049
361 East Broad Street
Columbus, OH 43216
   (614) 466-6422
   Harold Kohn

Ohio
Environmental Protection Agency
Public Wastewater
Box 1049
361 East Broad Street
Columbus, OH 43216
   (614) 466-5363
   Gene Wright

Ohio
Environmental Protection Agency
Solid Waste
P.O. Box 1049
Columbus, OH 43216
   (614) 466-8934
   Coordinator: Donald Day

Ohio River Valley Water Sanitation
   Commission
414 Walnut Street
Ninth Floor
Cincinnati, OH 45202
   (513) 421-1151
   Executive Director: Leo Weaver

Oklahoma
State Department of Health
Industrial and Solid Waste Division
N.E. 10th and Stonewall
Oklahoma City, OK 73105
   (405) 271-5338
   Director: H.A. Caves

Oregon
Department of Environmental Quality
Solid Waste
522 S.W. Fifth Avenue
P.O. Box 1760
Portland, OR 97207
   (503) 229-6403
   Coordinator: Bob Brown

Pennsylvania
Department of Environmental Resources
Division of Solid Waste Management
P.O. Box 2063
Harrisburg, PA 17120
 (717) 787-7381
 Director: Donald Lazarchik

Rhode Island
Department of Environmental Manage-
 ment
Division of Solid Waste Management
204 Health Building
Davis Street
Providence, RI 02908
 (401) 277-2808
 Supervisor: John Auinn, Jr.

Solar Energy Institute of America
1110 Sixth Street, N.W
Washington, DC 20001
 (202) 667-6611
 Vice President: Luana Moore

South Carolina
Bureau of Wastewater and Stream
 Quality Control
Industrial and Agricultural Waste-
 water Division
2600 Bull Street
Columbia, SC 29201
 (803) 758-5483
 Director: Robert Gross

South Carolina
Department of Health and Environmen-
 tal Control
Solid Waste Management Division
2600 Bull Street
Columbia, SC 29201
 (803) 758-5681
 Director: Hartsill Truesdale

Tennessee
Department of Public Health
Division of Sanitation and Solid
 Waste Management
Solid Waste Management Section
320 Capitol Hill Building
Nashville, TN 37219
 (615) 741-3424
 Tom Tiesler

Texas
Department of Health
Division of Wastewater Technology and
 Surveillance
1100 West 49th Street
Austin, TX 78756
 (512) 458-7111
 Director: Henry Dabney

Texas
Department of Water Resources
Solid Waste Branch
Box 13246
Capitol Station
Austin, TX 78711
 (512) 475-2041
 Chief: Jay Snow

U.S.
Department of Agriculture
Science and Education Department
Energy Research
Washington, DC 20250
 (301) 344-2740
 Coordinator: Dr. Altman

U.S.
Department of Energy
Alaska Power Administration
Federal Building
P.O. Box 50
Juneau, AK 99802
 (907) 586-7405

U.S.
Department of Energy
Bonneville Power Administration
1002 N.E. Holladay Street
P.O. Box 3621
Portland, OR 97208
 (503) 234-3361

U.S.
Department of Energy
Conservation and Solar Applications
Washington, DC 20545
 (202) 376-4000
 Assistant Secretary: Omi Walden

U.S.
Department of Energy
Director of Administration

Washington, DC 20545
  (202) 376-4000
  Director: William Heffelfinger

U.S.
Department of Energy
Energy Information Administration
Washington, DC 20545
  (202) 376-4000
  Administrator: Lincoln Moses

U.S.
Department of Energy
Federal Energy Regulatory Commission
Atlanta Regional Office
730 Peachtree Street
Atlanta, GA 30308
  (404) 257-4134
  Regional Representative

U.S.
Department of Energy
Federal Energy Regulatory Commission
Chicago Regional Office
230 South Dearborn Street
Chicago, IL 60604
  (312) 353-6171
  Regional Representative

U.S.
Department of Energy
Federal Energy Regulatory Commission
Electric Power Regulation Offices
825 North Capitol Street, N.E.
Washington, DC 20426
  (202) 275-4875
  Chief of Environmental Analysis

U.S.
Department of Energy
Federal Energy Regulatory Commission
Environmental Analysis Branch
825 North Capitol Street, N.E.
Washington, DC 20426
  (202) 275-4875
  Chief: Quentin Adson

U.S.
Department of Energy
Federal Energy Regulatory Commission
Fort Worth Regional Office
819 Taylor Street

Fort Worth, TX 76102
  (817) 334-2631
  Regional Representative

U.S.
Department of Energy
Federal Energy Regulatory Commission
New York Regional Office
26 Federal Plaza
New York, NY 10007
  (212) 264-3687
  Regional Representative

U.S.
Department of Energy
Federal Energy Regulatory Commission
Regulatory Analysis Office
825 North Capitol Street, N.E.
Washington, DC 20426
  (202) 275-6569
  Dr. Jack Heinemann

U.S.
Department of Energy
Federal Energy Regulatory Commission
San Francisco Regional Office
555 Battery Street
San Francisco, CA 94111
  (415) 556-3581
  Regional Representative

U.S.
Department of Energy
NEPA Affairs Division
Mail Station 4G-064
Forrestal Building
Washington, DC 20585
  (202) 252-4600
  Acting Director: Dr. Robert Stern

U.S.
Department of Energy
Region I
150 Causeway Street
Room 700
Boston, MA 02114
  (617) 223-3705
  Director

U.S.
Department of Energy
Region II

20 Federal Plaza
Room 3200
New York, NY   10007
   (212) 264-1040
   Director

U.S.
Department of Energy
Region III
1421 Cherry Street
Philadelphia, PA   19102
   (215) 597-3870
   Director

U.S.
Department of Energy
Region IV
1655 Peachtree Street, N.E.
Atlanta, GA   30309
   (404) 881-2661
   Director

U.S.
Department of Energy
Region V
175 West Jackson Boulevard
Room A-333
Chicago, IL   60604
   (312) 353-8770
   Director

U.S.
Department of Energy
Region VI
P.O. Box 35228
Dallas, TX   75235
   (214) 749-7626
   Director

U.S.
Department of Energy
Region VII
324 East 11th Street
Kansas City, MO   64152
   (816) 374-5936
   Director

U.S.
Department of Energy
Region VIII
Belmar Branch
P.O. Box 26247

Lakewood, CO   80226
   (303) 234-3195
   Director

U.S.
Department of Energy
Region IX
111 Pine Street
Third Floor
San Francisco, CA   94111
   (415) 556-7220
   Director

U.S.
Department of Energy
Region X
1990 Federal Building
915 Second Avenue
Seattle, WA   98174
   (206) 442-7320
   Director

U.S.
Department of Energy
Southeastern Power Administration
Elberton, GA   30635
   (404) ___ 3261
   Director

U.S.
Department of Energy
Southwestern Power Administration
Page Belcher Federal Building
P.O. Drawer 1619
Tulsa, OK   74101
   (918) 581-7474
   Director

U.S.
Department of Energy
Western Regional Office
P.O. Box 3402
Golden, CO   80401
   (303) 234-7500
   Director

U.S.
Department of the Interior
Energy and Resources
C Street Between 18th and 19th
   Streets, N.W.
Washington, DC   20240

(202) 343-1100
  Associate Solicitor: John Leshy

U.S.
Department of the Interior
Fish and Wildlife Service
National Power Plant Team
1451 Green Road
Ann Arbor, MI 48105
  (313) 668-2365
  Director

U.S.
Department of the Interior
Office of Environmental Project
  Review
Energy Facilities
Office of the Secretary
Washington, DC 20240
  (202) 343-6128
  Staff: Lillian Stone

U.S.
Energy Coordinating Committee
7A123 Forrestal Building
Washington, DC 20314
  (202) 252-5421

U.S.
Environmental Protection Agency
Solid Waste
401 M Street, S.W.
Washington, DC 20460
  (202) 755-2673
  Deputy Assistant Administrator:
    Steffen Plehn

U.S.
Environmental Protection Agency
Water and Waste Management
401 M Street, S.W.
Washington, DC 20460
  (202) 755-2673
  Thomas Jorlinq

U.S.
Federal Power Commission
Office of Energy Systems
825 North Capitol Street, N.E.
Room 4306
Washington, DC 20426
  (202) 386-3335

U.S.
Nuclear Regulatory Commission
Division of Engineering
Environmental Technology Group
Washington, DC 20555
  (301) 492-7017
  Assistant Director: Daniel Muller

U.S.
Nuclear Regulatory Commission
Division of Engineering
Environmental Technology Group
Aquatic Resources Section
Washington, DC 20555
  (301) 492-8208
  Section Leader: Robert Samworth

U.S.
Nuclear Regulatory Commission
Division of Engineering
Environmental Technology Group
Environmental Engineering Branch
Washington, DC 20555
  (301) 492-8448
  Branch Chief: Ronald Ballard

U.S.
Nuclear Regulatory Commission
Division of Engineering
Environmental Technology Group
Regional Impact Analysis Section
Washington, DC 20555
  (301) 492-8556
  Section Leader: Donald Cleary

U.S.
Nuclear Regulatory Commission
Division of Engineering
Environmental Technology Group
Site Analysis Section
Washington, DC 20555
  (301) 492-7856
  Section Leader: Len Soffer

U.S.
Nuclear Regulatory Commission
Division of Engineering
Environmental Technology Group
Terrestrial Resources Section
Washington, DC 20555
  (301) 492-8251
  Section Leader: Jerry Kline

U.S.
Nuclear Regulatory Commission
Division of Engineering
Environmental Technology Group
Utility Finance Branch
Washington, DC 20555
  (301) 492-8336
  Branch Chief: Jerome Saltzman

U.S.
Nuclear Regulatory Commission
Division of Engineering
Environmental Technology Group
Utility Section
Washington, DC 20555
  (301) 492-8593
  Section Leader: Darrel Nash

U.S.
Nuclear Regulatory Commission
Division of Site Safety and Environ-
  mental Analysis
Washington, DC 20555
  (301) 492-7000
  Director: Richard DeYoung

U.S.
Nuclear Regulatory Commission
Division of Site Safety and Environ-
  mental Analysis
Effluent Treatment Systems Branch
Washington, DC 20555
  (301) 492-8361
  Chief: John Collins

U.S.
Nuclear Regulatory Commission
Environmental Projects
P-518
Washington, DC 20555
  (301) 492-8446
  Assistant Director: Voss Moore

U.S.
Nuclear Regulatory Commission
Region I
631 Park Avenue
King of Prussia, PA 19406
  (215) 337-1150
  Director: Boyce Grier

U.S.
Nuclear Regulatory Commission
Region II
Atlanta, GA 30303
  (404) 221-4503
  Director: James O'Reilly

U.S.
Nuclear Regulatory Commission
Region III
799 Roosevelt Road
Glen Ellyn, IL 60137
  (312) 932-2611
  Chief: Thomas Essig

U.S.
Nuclear Regulatory Commission
Region IV
611 Ryan Plaza Drive
Arlington, TX 76011
  (817) 334-2841
  Director: Karl Seyfrit

U.S.
Nuclear Regulatory Commission
Region V
1990 North California Boulevard
Walnut Creek, CA 94596
  (415) 486-3141
  Director: Robert Engelken

Utah
Environmental Coordinating Committee
Division of Oil, Gas, and Mining
124 State Capitol
Salt Lake City, UT 84111
  (801) 533-5794
  Ron Daniels

Utah
Environmental Coordinating Committee
State Energy Office
124 State Capitol
Salt Lake City, UT 84111
  (801) 533-5794
  Buzz Hunt

Utah
State Division of Health
Bureau of Solid Waste Management
44 Medical Drive

Salt Lake City, UT   84113
   (801) 533-6111
   Director: Dale Parker

Virginia
Division of Solid and Hazardous Waste
   Management
Madison Building
Richmond, VA   23219
   (804) 786-5271
   Director: William Gilley

Virginia
Energy Resource Advisory Commission
823 East Main Street
Room 300
Richmond, VA   23219
   (804) 786-8451
   Executive Secretary: Louis Lawson

Virginia
State Corporation Commission
Division of Energy Regulations
P.O. Box 1197
Blanton Building
9th Floor
Richmond, VA   23219
   (804) 786-3614
   Director: James Whittine

Virginia
State Water Control Board
Bureau of Applied Technology
2111 North Hamilton Street
Richmond, VA   23230
   (804) 257-0056
   Director: Louis Lawson

Virgin Islands
Department of Public Works
Division of Utilities and Sanitation
St. Thomas, VI   00801
   (809) 774-2070
   Commissioner: Gordon Finch

Washington
Department of Ecology
Solid Waste Management Division
Olympia, WA   98504
   (206) 753-2800
   Division Supervisor: Earl Tower

Washington
Department of Social and Health
   Services
Water Supply and Waste Unit
M/S LD-11
Olympia, WA   98504
   (206) 753-5954
   Supervisor: James Pluntze

West Virginia
State Department of Health
Solid Waste Program
1800 East Washington Street
Charleston, WV   25305
   (304) 348-2987
   Director: Dale Parsons

West Virginia Surface Mining and
   Reclamation Association
Office of the President
1624 Kanawha Boulevard East
Charleston, WV   25311
   (304) 346-5318
   President: Ben Lusk

Wisconsin
Department of Natural Resources
Division of Environmental Standards
Solid Waste Management Section
P.O. Box 7921
Madison, WI   53701
   (608) 266-1327
   Director: Robert Krill

Wisconsin
Department of Natural Resources
Division of Environmental Studies
Industrial Waste Water Section
Madison, WI   53701
   (608) 266-0289
   Chief: Paul Didier

Wyoming
Department of Environmental Quality
Solid Waste Program
Hathaway Building
Cheyenne, WY   82002
   (307) 328-3752
   Supervisor: Charles Porter

10. HEALTH-RELATED AGENCIES AND ORGANIZATIONS

American Medical Association
Environmental, Public, and Occupational Health
1535 North Dearborn
Chicago, IL 60610
  (312) 751-6532
  Assistant Director: Ward Duel

American Public Health Association
Office of the Executive Director
1015 18th Street, N.W.
Washington, DC 20036
  (202) 467-5000
  Dr. William McBeath

Arizona
Department of Health Services
Division of Environmental Health
  Services
1740 West Adams Street
Phoenix, AZ 85007
  (602) 255-1024
  Assistant Director: Bruce Scott

Association of State and Territorial
  Health Officials
101 Second Street, N.E.
Washington, DC 20002
  (202) 547-3470
  President: Eugene Fowinkle

California
Department of Health
714 P Street
Sacramento, CA 95814

  (916) 445-0498
  Linda Bates

District of Columbia
Department of Environmental Services
Environmental Health Administration
415 12th Street, N.W.
Room 301
Washington, DC 20004
  (202) 724-4113
  Chief: Angelos Tompros

Hawaii
Department of Health
Environmental Programs
Box 3378
Honolulu, HI 96801
  (808) 548-4139
  Deputy Director: James Kumagai

Louisiana
Health and Human Resources Admin-
  istration
Bureau of Environmental Health
P.O. Box 60630
New Orleans, LA 70160
  (504) 568-5100
  Director: James Coerver

Michigan
Department of Public Health
3500 North Logan Street
Box 30035
Lansing, MI 48909
  (517) 373-1320

Director: Maurice Reizen

Mississippi
State Board of Health
P.O. Box 1700
Jackson, MS 39205
  (601) 354-6616
  Director of Environmental Health:
    Joe Brown

National Environmental Health
  Association
Office of the Director
1200 Lincoln
Suite 704
Denver, CO 80203
  (303) 861-9090
  Executive Director: Dr. L.J. Krone

Oak Ridge National Laboratory
Health and Environmental Studies
  Program
P.O. Box X
Oak Ridge, TN 37830
  (615) 483-8611
  Director: Anna Hammons

Texas
Department of Health
Bureau of Environmental Health
1100 West 49th Street
Austin, TX 78756
  (512) 458-7111
  Chief: David Cochran

Texas
Department of Health
Environmental and Consumer Health
  Protection
1100 West 49th Street
Austin, TX 78756
  (512) 458-7111
  Deputy Commissioner: G.R. Herzik,
    Jr.

U.S.
Department of HEW
Division of Environmental Health
  Services
Center for Disease Control
Atlanta, GA 30333

  (404) 633-3311
  Deputy Director: Dr. Frank Lisella

U.S.
Department of HEW
Division of Procurement Services
Health Car Financing Administration
1710 Gwynn Oak Avenue
Area D-2
Baltimore, MD 21207
  Acting Director: Paul Fiore

U.S.
Department of HEW
National Institute of Environmental
  Health Sciences
9000 Rockville Pike
Bethesda, MD 20205
  (301) 496-4000
  Director: David Rall

U.S.
Department of HEW
Public Health Service
Office of the Assistant Secretary
  for Health
Environmental Affairs
200 Independence Avenue, S.W.
Washington, DC 20201
  (202) 245-6296
  Senior Advisor: James Dickson, III

U.S.
Department of the Army
The Surgeon General
HQDA (DASG-PSP)
Washington, DC 20310
  (202) 697-2796
  Lee Herwig, COL

U.S.
Environmental Protection Agency
Health and Ecological Effects
401 M Street, S.W.
Washington, DC 20460
  (202) 755-2673
  Deputy Assistant Administrator:
    William Murray

Utah
Environmental Coordinating Committee

Division of Health
Branch of Environmental Health
124 State Capitol
Salt Lake City, UT 84111
  (801) 533-5794
  Dennis Dalley

Utah
State Division of Health
Bureau of Environmental Epidemiology
44 Medical Drive
Salt Lake City, UT 84113
  (801) 533-6111
  Director: Wanlass Southwick

Virginia
Department of Health
Commission for the Environment
927 Madison Building
Richmond, VA 23219
  (804) 786-6277
  Deputy Assistant: Eric Bartsch

11. HOUSING-RELATED AGENCIES AND ORGANIZATIONS

California
Department of Housing and Community
  Development
921 10th Street
Sixth Floor
Sacramento, CA 95814
  (916) 445-4725
  Dave Williamson

Connecticut
Housing Financing Authority
190 Trumbull Street
Hartford, CT 06103
  (203) 525-9311
  Director

Hawaii
Social Services and Housing Depart-
  ment
Housing Authority
1390 Miller Street
Honolulu, HI 96813
  (808) 848-3211
  Director

Los Angeles City
Housing Authority
515 Columbia Avenue
Los Angeles, CA 90017
  (213) 483-6440
  Director

Maryland
Economic and Community Development
  Department

Housing
2525 Riva Road
Annapolis, MD 21401
  (301) 269-3524
  Director

Minnesota
Housing
Hanover Building
480 Cedar Street
St. Paul, MN 55101
  (612) 296-7608
  Director

New York City
Housing Authority
250 Broadway
New York, NY 10007
  (212) 433-2525
  Director

New York
Housing and Community Renewal Division
Alfred Smith State Office Building
Albany, NY 12225
  (518) 474-8580
  Director

U.S.
Department of HUD
Atlanta Field Office
1371 Peachtree Street, N.E.
Atlanta, GA 30309
  (404) 881-4585
  Officer in Charge: Russell Marane

U.S.
Department of HUD
Boston Field Office
Kennedy Federal Building
Boston, MA 02203
  (617) 223-4066
  Officer in Charge: Edward Martin

U.S.
Department of HUD
Chicago Field Office
300 South Wacker Drive
Chicago, IL 60606
  (312) 353-5680
  Officer in Charge: Ronald Gatton

U.S.
Department of HUD
Community Planning and Development
451 Seventh Street, S.W.
Washington, DC 20410
  (202) 755-6228
  Director

U.S.
Department of HUD
Dallas Field Office
Cabell Federal Building
1100 Commerce Street
Dallas, TX 75242
  (214) 749-7401
  Officer in Charge: Thomas Armstrong

U.S.
Department of HUD
Denver Field Office
Executive Towers
1405 Curtis Street
Denver, CO 80202
  (303) 837-4513
  Officer in Charge: Betty Miller

U.S.
Department of HUD
Kansas City Field Office
300 Federal Office Building
911 Walnut Street
Kansas City, MO 64106
  (816) 374-2661
  Officer in Charge: William
    Anderson

U.S.
Department of HUD
Office of Environmental Quality
Room 7274
451 Seventh Street, S.W.
Washington, DC 20410
  (202) 755-6300
  Director: Richard Broun

U.S.
Department of HUD
Office of Housing
451 Seventh Street, S.W.
Washington, DC 20410
  (202) 755-5658
  Director

U.S.
Department of HUD
Office of Multifamily Housing Develop-
  ment
451 Seventh Street, S.W.
Washington, DC 20410
  (202) 755-5111
  Director: George Hipps, Jr.

U.S.
Department of HUD
Office of Single Family Housing
451 Seventh Street, S.W.
Washington, DC 20410
  (202) 755-5111
  Director: Milton Francis

U.S.
Department of HUD
Philadelphia Field Office
Curtis Building
6th and Walnut Streets
Philadelphia, PA 19106
  (215) 597-2560
  Officer in Charge: Thomas Maloney

U.S.
Department of HUD
Program Information Center
451 Seventh Street, S.W.
Room 1104
Washington, DC 20410
  (202) 755-6420
  Louise North

<u>U.S.</u>
Department of HUD
San Francisco Field Office
450 Golden Gate Avenue
P.O. Box 36003
San Francisco, CA   94102
  (415) 556-4752
  Officer in Charge: Emma McFarlin

<u>U.S.</u>
Department of HUD
Seattle Field Office
3003 Arcade Plaza Building
1321 Second Avenue
Seattle, WA   98101
  (206) 442-5414
  Officer in Charge: George Roybal

<u>Virginia</u>
Housing Development Authority
111 South Sixth Street
Richmond, VA   23219
  (804) 786-8241
  Director

<u>Wisconsin</u>
Housing Financial Authority
131 West Wilson Street
Madison, WI   53703
  (608) 266-7884
  Director

12. POPULATION-RELATED AGENCIES AND ORGANIZATIONS

Population Reference Bureau
1337 Connecticut Avenue, N.W.
Washington, DC 20036
   (202) 785-4664
   President: Robert Avedon

U.S.
Department of Commerce
Bureau of the Census
Atlanta Regional Office
1365 Peachtree Street, N.E.
Atlanta, GA 30309
   (404) 881-2271
   Director

U.S.
Department of Commerce
Bureau of the Census
Boston Regional Office
441 Stuart Street
Boston, MA 02116
   (617) 223-2327
   Director

U.S.
Department of Commerce
Bureau of the Census
Census Data Preparation Division
1201 East Tenth Street
Building 48
Jeffersonville, IN 47132
   (812) 283-3511
   Director

U.S.
Department of Commerce
Bureau of the Census
Charlotte Regional Office
230 South Tyron Street
Charlotte, NC 28202
   (704) 372-7351
   Director

U.S.
Department of Commerce
Bureau of the Census
Chicago Regional Office
55 East Jackson Boulevard
Chicago, IL 60604
   (312) 353-4932
   Director

U.S.
Department of Commerce
Bureau of the Census
Dallas Regional Office
1100 Commerce Street
Dallas, TX 75242
   (214) 749-2814
   Director

U.S.
Department of Commerce
Bureau of the Census
Denver Regional Office
575 Union Boulevard
Denver, CO 80225
   (303) 234-3924
   Director

U.S.
Department of Commerce
Bureau of the Census
Detroit Regional Office
231 West Lafayette
Detroit, MI 48226
  (313) 226-7742
  Director

U.S.
Department of Commerce
Bureau of the Census
Information Office
Washington, DC 20233
  (301) 763-7273
  Director

U.S.
Department of Commerce
Bureau of the Census
Kansas City Regional Office
4th and State Streets
Kansas City, MO 66101
  (816) 374-4601
  Director

U.S.
Department of Commerce
Bureau of the Census
Los Angeles Regional Office
11777 San Vicente Boulevard
Los Angeles, CA 90049
  (213) 824-7317
  Director

U.S.
Department of Commerce
Bureau of the Census
New York Regional Office
26 Federal Plaza
New York, NY 10007
  (212) 264-3860
  Director

U.S.
Department of Commerce
Bureau of the Census
Philadelphia Regional Office
600 Arch Street
Philadelphia, PA 19106
  (215) 597-4920
  Director

U.S.
Department of Commerce
Bureau of the Census
Seattle Regional Office
1700 Westlake Avenue, North
Seattle, WA 98109
  (206) 442-7800
  Director

U.S.
Department of Labor
Office of Information, Publications,
  and Reports
200 Constitution Avenue, N.W.
Washington, DC 20210
  (202) 523-8165
  Director: John Leslie

U.S.
Department of Labor
Region I
Kennedy Federal Building
Boston, MA 02203
  (617) 223-5430
  Regional Information Director:
    Paul Neal

U.S.
Department of Labor
Region II
1515 Broadway
New York, NY 10036
  (212) 399-5252
  Regional Information Director:
    Ed Weintraub

U.S.
Department of Labor
Region III
3535 Market Street
Philadelphia, PA 19101
  (215) 596-1116
  Regional Information Director:
    Jack Hord

U.S.
Department of Labor
Region IV
1371 Peachtree Street, N.E.
Atlanta, GA 30309
  (404) 526-5366

Regional Information Director:
   Frances Ridgway

U.S.
Department of Labor
Region V
230 South Dearborn Street
Chicago, IL 60606
   (312) 353-4122
   Regional Information Director:
   John Mellott

U.S.
Department of Labor
Region VI
555 Griffin Square Building
Dallas, TX 75202
   (214) 749-3842
   Regional Information Director:
   Les Gaddie

U.S.
Department of Labor
Region VII
911 Walnut Street
Kansas City, MO 64106
   (816) 374-5941
   Regional Information Director:
   Patrick Hand

U.S.
Department of Labor
Region VIII
1961 Stout Street
Denver, CO 80294
   (303) 837-3791
   Regional Information Director:
   Ernest Sanchez

U.S.
Department of Labor
Region IX
450 Golden Gate Avenue
San Francisco, CA 94102
   (415) 556-8754
   Regional Information Director:
   Joe Kirkbride

U.S.
Department of Labor
Region X

909 First Avenue
Seattle, WA 98174
   (206) 442-1545
   Regional Information Director:
   Jack Strickland

U.S.
Department of the Interior
Bureau of Indian Affairs
Washington, DC 20240
   (202) 343-7445
   Public Information Staff

13. RECREATION-RELATED AGENCIES AND ORGANIZATIONS

Arizona
Outdoor Recreation Coordinating
  Commission
1333 West Camelback Road
Suite 206
Phoenix, AZ 85013
  (602) 271-5013
  Recreation Planning Coordinator:
  Lyle Bair

California
Department of Parks and Recreation
1200 K Street Mall
Sacramento, CA 95814
  (916) 322-2481
  Clark Woy

Connecticut
Department of Environmental Pro-
  tection
Parks and Recreation Unit
State Office Building
165 Capitol Avenue
Hartford, CT 06115
  (203) 566-5599
  Chief: William Miller

Connecticut Forest and Park
  Association
1010 Main Street
P.O. Box 389
East Hartford, CT 06108
  (203) 289-3637
  President: David Smith

Delaware
Department of Natural Resources and
  Environmental Control
Division of Parks and Recreation
Tatnall Building
P.O. Box 1401
Dover, DE 19901
  (302) 678-4401
  Director: John Wilson, III

Georgia
Department of Natural Resources
Recreation Planning
270 Washington Street, S.W.
Atlanta, GA 30334
  (404) 656-3530
  Chief: Robin Jackson

Hawaii
Department of Land and Natural
  Resources
Division of State Parks
P.O. Box 621
Honolulu, HI 96809
  (808) 548-7455
  Administrator: Joseph Souza, Jr.

Idaho Recreation and Park Society
6709 East Powerline
Nampa, ID 83651
  (208) 384-2284
  President: Donald Denton

Indiana
Department of Natural Resources

Division of Outdoor Recreation
608 State Office Building
Indianapolis, IN 46204
  (317) 633-6344
  Head: Gerald Pagac

Indiana
Department of Natural Resources
Division of State Parks
608 State Office Building
Indianapolis, IN 46204
  (317) 633-6344
  Head: William Walters

Los Angeles City
Recreation and Parks Department
City Hall
East 200 Spring Street
Room 1330
Los Angeles, CA 90012
  (213) 485-5508
  Director

Maryland
Department of Natural Resources
Park Service
Tawes State Office Building
Annapolis, MD 21401
  (301) 269-3761
  Director: William Parr

Massachusetts
Executive Office of Environmental
  Affairs
Department of Metropolitan District
  Commission
Division of Parks
20 Sommerset Street
Boston, MA 02108
  (617) 727-5114
  Director: Robert Williams

Massachusetts Forest and Park
  Association
Three Joy Street
Boston, MA 02108
  (617) 742-2553
  President: Herbert Pratt

Michigan
Department of Natural Resources
Parks Division

Box 30028
Lansing, MI 48909
  (517) 373-1220
  Chief: Jack Butterfield

Minnesota
Department of Natural Resources
Division of Parks
300 Centennial Building
658 Cedar Street
St. Paul, MN 55155
  (612) 296-2270
  Director: Donald Davison

Mississippi
Park Commission
717 Robert E. Lee Building
Jackson, MS 39201
  (601) 354-6321
  Executive Director: T.P. Edwards

Montana
Department of Fish and Game
Parks
1420 East Sixth
Helena, MT 59601
  (406) 449-2335
  Administrator: Ron Holliday

National Association of State Outdoor
  Recreation Liason Officers
State Parks Superintendant
555 Trade Street, S.E.
Salem, OR 97310
  (503) 378-6305
  President: Dave Talbot

National Association of State Park
  Directors
Division of Recreation and Parks
Crown Building
202 Blount Street
Tallahassee, FL 32301
  President: Ney Landrum

National Parks and Conservation
  Association
1701 18th Street, N.W.
Washington, DC 20009
  (202) 265-2717
  President: Anthony Smith

National Recreation and Park
  Association
1601 North Kent Street
Arlington, VA 22209
  (703) 525-0606
  President: David Ladlaw

Nevada
Department of Conservation and
  Natural Resources
Division of State Parks
1923 North Carson Street
Suite 210
Carson City, NV 89710
  (702) 885-4384
    Deputy Administrator: John
      Richardson

New Jersey
Department of Environmental Conser-
  vation
Division of Parks and Forestry
P.O. Box 1420
Trenton, NJ 08625
  (609) 292-2733
  Director: Alfred Guido

New Jersey
Department of Environmental Protec-
  tion
Green Acres
P.O. Box 1390
Trenton, NJ 08625
  (609) 292-0745
  Chief: Robert Stokes

New York City
Parks and Recreation Department
Arsenal Building
830 Fifth Avenue
New York, NY 10021
  (212) 360-8111
  Director

New York
State Office of Parks and Recreation
Empire State Plaza
Albany, NY 12238
  (518) 474-0456
   Director of Environmental Manage-
     ment: Dr. Peter Buttner

North Carolina
Department of Natural Resources and
  Community Development
Recreation
P.O. Box 27687
Raleigh, NC 27611
  (919) 733-4984
  Jim Stevens

Ohio Parks and Recreation Association
33 South James Road
Suite 233
Columbus, OH 43213
  (614) 231-0781
  President: Gary Fenton

South Dakota
Wildlife, Parks, and Forestry Depart-
  ment
Parks and Recreation Division
Sigurd Anderson Building
Pierre, SD 57501
  (605) 224-3391
  Director: Lowen Schuett

Texas
Parks and Wildlife Department
4200 Smith School Road
Austin, TX 78744
  (512) 475-4888
  Executive Director: Henry Burkett

Texas
Parks and Wildlife Department
Parks
4200 Smith School Road
Austin, TX 78744
  (512) 475-4888
  Director: Paul Schlimper

U.S.
Department of Agriculture
Soil Conservation Service
P.O. Box 2890
Washington, DC 20013
  (202) 447-3921
  Recreation Specialist: Ralph Wilson

U.S.
Department of Agriculture
Soil Conservation Service

Nebraska Region
100 Centennial Mall North
Lincoln, NE 68508
   Recreation Specialist: C.V. Bohart

U.S.
Department of Agriculture
Soil Conservation Service
Oregon Region
Federal Building
Room 507
511 N.W. Broadway
Portland, OR 97204
   Recreation Specialist: Clarence
      Maesner

U.S.
Department of Agriculture
Soil Conservation Service
Pennsylvania Region
1974 Sproul Road
Broomall, PA 19008
   Recreation Specialist: Hans Uhlig

U.S.
Department of Agriculture
Soil Conservation Service
Texas Region
P.O. Box 6567
Fort Worth, TX 76115
   Recreation Specialist: Ross Miller

U.S.
Department of the Interior
Bureau of Outdoor Recreation
Office of Environmental Affairs
Washington, DC 20240
   (202) 343-1005
   Chief: Louis Reid, Jr.

U.S.
Department of the Interior
National Park Service
Mid-Atlantic Regional Office
143 South Third Street
Philadelphia, PA 19106
   (215) 597-7013
   Director

U.S.
Department of the Interior
National Park Service

Midwest Regional Office
1709 Jackson Street
Omaha, NE 68102
   (402) 864-3431
   Director

U.S.
Department of the Interior
National Park Service
National Capital Regional Office
1100 Ohio Drive, S.W.
Washington, DC 20242
   (202) 426-6612
   Director

U.S.
Department of the Interior
National Park Service
North Atlantic Regional Office
150 Causeway Street
Boston, MA 02109
   (617) 223-3788
   Director

U.S.
Department of the Interior
National Park Service
Office of Planning and Environmental
   Quality
18th and C Streets, N.W.
Washington, DC 20240
   (202) 343-5625
   Director

U.S.
Department of the Interior
National Park Service
Pacific Northwest Regional Office
601 Fourth and Pike Building
Seattle, WA 98101
   (206) 399-5565
   Director

U.S.
Department of the Interior
National Park Service
Rocky Mountain Regional Office
P.O. Box 25287
Denver, CO 80225
   (303) 234-2500
   Director

U.S.
Department of the Interior
National Park Service
Southeast Regional Office
1895 Phoenix Boulevard
Atlanta, GA 30349
  (404) 289-7594
  Director

U.S.
Department of the Interior
National Park Service
Southwest Regional Office
Box 728
Santa Fe, NM 87501
  (505) 476-3388
  Director

U.S.
Department of the Interior
National Park Service
Western Regional Office
450 Golden Gate Avenue
San Francisco, CA 94102
  (415) 556-4196
  Director

Utah
Division of Parks and Recreation
Environmental Coordinating Committee
124 State Capitol
Salt Lake City, UT 84111
  (801) 533-5794
  Ken Travous

Utah
State Department of Natural Resources
Division of Parks and Recreation
1596 W.N. Temple
Salt Lake City, UT 84116
  (801) 533-6011
  Parks Planning Supervisor: Lionel
    Brown

Utah
State Department of Natural Resources
Division of Parks and Recreation
Parks Environment and Special Studies
1596 W.N. Temple
Salt Lake City, UT 84116
  (801) 533-6011
  Coordinator: Max Jensen

Utah
State Department of Natural Resources
Outdoor Recreation Agency
807 E.S. Temple
Suite 101
Salt Lake City, UT 84102
  (801) 533-5691
  Project Director: Ross Elliott

Vermont
Agency of Environmental Conservation
Parks and Recreation
Montpelier, VT 05602
  (802) 828-3375
  Rodney Barber

Virginia
Commission of Outdoor Recreation
8th Street Office Building
803 East Broad Street
Richmond, VA 23219
  (804) 786-2036
  Director: Rob Blackmore

Virginia
Department of Conservation and
  Economic Development
Division of State Parks
1201 State Office Building
Richmond, VA 23219
  (804) 786-2132
  Commissioner: Ben Bolen

Virginia
Northern Virginia Regional Park
  Authority
11001 Popes Head Road
Fairfax, VA 22030
  (703) 278-8880
  Executive Director: Darrell Winslow

Washington
Department of Natural Resources
Recreation
Public Lands Building
Olympia, WA 98504
  (206) 753-5327
  Division Supervisor: Al O'Donnell

Washington
State Parks and Recreation Commission
7150 Clearwater Lane

Olympia, WA 98504
  (206) 753-5755
  Director: Charles Odegaard

<u>Washington</u>
State Parks and Recreation Commission
Environmental Coordination
7150 Clearwater Lane
Olympia, WA 98504
  (206) 753-5755
  Chief: David Heiser

<u>Wisconsin</u>
Department of Natural Resources
Bureau of Parks and Recreation
Box 7921
Madison, WI 53707
  (608) 266-2152
  Director: Donald Mackie

<u>Wisconsin Park and Recreation
  Association</u>
7600 West North Avenue
Wauwatosa, WI 53213
  (414) 258-3140
  President: Augie Revoy

14. TRANSPORTATION-RELATED AGENCIES AND ORGANIZATIONS

Airport Operators Council International
Environment Office
1700 K Street, N.W.
Washington, DC 20006
  (202) 296-3270
  Director of Environmental Programs

Air Transport Association of America
1709 New York Avenue, N.W.
Washington, DC 20006
  (202) 872-4000
  Director

American Association of State Highway and Transportation Officials
444 North Capitol Street, N.W.
Suite 225
Washington, DC 20001
  (202) 624-5800
  Executive Director

American Public Transit Association
1100 17th Street, N.W.
Washington, DC 20036
  (202) 331-1100
  Executive Vice President:
    B.R. Stokes

American Railway Engineering Association
59 East Van Buren Street
Chicago, IL 60605
  (312) 939-0780
  Director

American Road Builders Association
525 School Street, S.W.
Washington, DC 20024
  (202) 488-2722
  Raymond Crowe

American Society of Traffic and Transportation
547 West Jackson Boulevard
Chicago, IL 60606
  (312) 939-2491
  Director

Association of American Railroads
American Railroad Building
1920 L Street, N.W.
Washington, DC 20036
  (202) 293-4000
  Director

California
Caltrans
Aeronautics
1120 N Street
Sacramento, CA 95814
  (916) 322-9954
  Burd Miller

California
Caltrans
Planning
1120 N Street
Sacramento, CA 95814
  (916) 445-9025
  Darrel Husum

California
Transportation Department
1120 N Street
Sacramento, CA 95814
  (916) 445-4616
  Director

Connecticut
Transportation Department
24 Wolcott Hill Road
Wethersfield, CT 06109
  (203) 566-5280
  Director

Hawaii
Transportation Department
869 Punchbowl Street
Honolulu, HI 96813
  (808) 548-3205
  Director

Highway Action Coalition
1346 Connecticut Avenue, N.W.
Room 731
Washington, DC 20036
  (202) 833-1845
  Director

Indiana
Traffic Safety and Vehicle Inspection
  Department
215 North Senate
Indianapolis, IN 46204
  (317) 633-5870
  Director

Institute of Transportation Engineers
P.O. Box 9234
Arlington, VA 22209
  (703) 527-5277
  Director of Technical Affairs

Los Angeles City
Traffic Department
City Hall
200 North Spring Street
Room 1200
Los Angeles, CA 90012
  (213) 485-2265
  Director

Maryland

Transportation Department
P.O. Box 8755
BWI Airport
Baltimore, MD 21240
  (301) 768-9520
  Director

Massachusetts
Department of Public Works
Transportation, Planning, and Devel-
  opment
100 Nashua Street
Boston, MA 02114
  (617) 727-4800
  Director: Thomas Humphrey

Massachusetts
Transportation and Construction
  Department
Transportation
1 Ashburton Place
Boston, MA 02108
  (617) 727-7680
  Director

Minnesota
Transportation Department
Transportation Building
St. Paul, MN 55155
  (612) 296-3131
  Director

Montana
Highway Department Safety Unit
Sixth and Roberts
Helena, MT 59601
  (406) 449-2071
  Director

National Air Transportation Association
1156-15th Street, N.W.
Washington, DC 20005
  (202) 293-2550
  Director

National Parking Association
1101-17th Street, N.W.
Washington, DC 20036
  (202) 296-4336
  Director

National Research Council
Transportation Research Board
Highway Research Information Service
2101 Constitution Avenue, N.W.
Washington, DC 20418
  (202) 389-6358
  Manager: Arthur Mobley

National Research Council
Transportation Research Board
Maritime Research Information Service
2101 Constitution Avenue, N.W.
Washington, DC 20418
  (202) 389-6687
  Manager: Davis Mellor

National Research Council
Transportation Research Board
Railroad Research Information Service
2101 Constitution Avenue, N.W.
Washington, DC 20418
  (202) 389-6611
  Manager: Frederick Houser

New York City
Traffic Department
28-11 Queens Plaza North
Long Island, NY 11101
  (212) 361-8000
  Director

New York
Transportation Department
Campus Building 5
Albany, NY 12232
  (518) 457-6195
  Director

North Carolina
Transportation Department
Highway Building
Raleigh, NC 27611
  (919) 733-2520
  Director

South Dakota
Transportation Department
Transportation Building
Pierre, SD 57501
  (605) 773-3265
  Director

Traffic Engineering
Institute of Traffic Engineers
1815 Fort Meyer Drive
Suite 905
Arlington, VA 22209
  (804) 527-5277
  Director

Transportation Association of America
1101-17th Street, N.W.
Washington, DC 20036
  (202) 296-2470
  Director

Transportation Board
2101 Constitution Avenue, N.W.
Washington, DC 20418
  (202) 393-8100
  Director

U.S.
Civil Aeronautics Board
Office of the General Counsel
1825 Connecticut Avenue, N.W.
Washington, DC 20428
  (202) 673-5205
  Steve Rothenburg

U.S.
Department of the Interior
Office of Environmental Project
  Review
Transportation
Office of the Secretary
Washington, DC 20240
  (202) 343-7564
  Staff: Joseph Fromme

U.S.
Department of Transportation
Assistant Secretary for Systems
Trisnet Secretariat
400 Seventh Street, S.W.
Washington, DC 20590
  (202) 426-0975
  Barbara Aleshire

U.S.
Department of Transportation
Coast Guard
Office of Marine Environment and
  Systems

400 Seventh Street, S.W.
Washington, DC 20590
  (202) 426-4000
    Chief: Rear Adm. Wayne Caldwell

U.S.
Department of Transportation
Environmental Analysis Division
400 Seventh Street, S.W.
Washington, DC 20590
  (202) 426-4000
    Joseph Canny

U.S.
Department of Transportation
Environmental, Civil Rights, and
  General Law
400 Seventh Street, S.W.
Washington, DC 20590
  (202) 426-4000
    Assistant General Counsel: Barclay
    Weber

U.S.
Department of Transportation
Environmental Coordination Division
400 Seventh Street, S.W.
Washington, DC 20590
  (202) 426-4000
    Eugene Lehr

U.S.
Department of Transportation
Federal Aviation Administration
400 Seventh Street, S.W.
Washington, DC 20590
  (202) 426-4000
    Associate Administrator for
      Airports: Robert Aaronson

U.S.
Department of Transportation
Federal Aviation Administration
Environment and Energy
400 Seventh Street, S.W.
Washington, DC 20590
  (202) 426-4000
    Director: John Wesler

U.S.
Department of Transportation
Federal Highway Administration

400 Seventh Street, S.W.
Washington, DC 20590
  (202) 426-4000
    Administrator: Karl Bowers

U.S.
Department of Transportation
Federal Highway Administration
Environmental Policy
400 Seventh Street, S.W.
Washington, DC 20590
  (202) 426-4000
    Director: M. Lash

U.S.
Department of Transportation
Federal Highway Administration
Environmental Programs Division
400 Seventh Street, S.W.
Washington, DC 20590
  (202) 426-4000
    Chief: Leon Larson

U.S.
Department of Transportation
Federal Highway Administration
Highway Operations
400 Seventh Street, S.W.
Washington, DC 20590
  (202) 426-4000
    Director: Vacant

U.S.
Department of Transportation
Federal Highway Administration
Office of Management Systems
400 Seventh Street, S.W.
Washington, DC 20590
  (202) 426-0630
    Director

U.S.
Department of Transportation
Federal Highway Administration
Region I and II
O'Brian Federal Building
Room 729
Clinton Avenue and North Pearl Street
Albany, NY 12207
  (518) 562-6476
    Regional Administrator: R.E. Kirby

U.S.
Department of Transportation
Federal Highway Administration
Region III
31 Hopkins Plaza
Baltimore, MD 21201
  (301) 922-2361
  Regional Administrator: W.H. White

U.S.
Department of Transportation
Federal Highway Administration
Region IV
1720 Peachtree Road, N.W.
Atlanta, GA 30309
  (404) 257-4078
  Regional Administrator: James Lacy

U.S.
Department of Transportation
Federal Highway Administration
Region V
18209 Dixie Highway
Homewood, IL 60430
  (312) 380-6300
  Regional Administrator: Donald Trull

U.S.
Department of Transportation
Federal Highway Administration
Region VI
819 Taylor Street
Fort Worth, TX 76102
  (817) 334-3232
  Regional Administrator: W.S.
   Mendenhall, Jr.

U.S.
Department of Transportation
Federal Highway Administration
Region VII
P.O. Box 19715
Kansas City, MO 64141
  (816) 926-7563
  Regional Administrator: John Kemp

U.S.
Department of Transportation
Federal Highway Administration
Region VIII
Denver Federal Center
Building 40

Denver, CO 80225
  (303) 234-4051
  Regional Administrator: Daniel Watt

U.S.
Department of Transportation
Federal Highway Administration
Region IX
2 Embarcadero Center
Suite 530
San Francisco, CA 94111
  (415) 556-3951
  Regional Administrator: Frank Hawley

U.S.
Department of Transportation
Federal Highway Administration
Region X
222 S.W. Morrison Street
Portland, OR 97204
  (503) 423-2065
  Regional Administrator: Louis
   Lybecker

U.S.
Department of Transportation
Federal Highway Administration
Right-Of-Way and Environment
400 Seventh Street, S.W.
Washington, DC 20590
  (202) 426-0539
  Associate Administrator: J.M.
   O'Connor

U.S.
Department of Transportation
Federal Highway Administration
Traffic Operations
400 Seventh Street, S.W.
Washington, DC 20590
  (202) 426-4000
  Director: Marshall Jacks, Jr.

U.S.
Department of Transportation
Federal Railroad Administration
400 Seventh Street, S.W.
Washington, DC 20590
  (202) 426-4000
  Administrator: John Sullivan

U.S.

Department of Transportation
Federal Railroad Administration
Region I
Independence Building
Room 1020
434 Walnut Street
Philadelphia, PA   19106
  (215) 597-0750
    Regional Administrator: Wallace
    Holl

U.S.
Department of Transportation
Federal Railroad Administration
Region II
1568 Willingham Drive
Suite 216-B
College Park, GA   30337
  (404) 264-7801
    Regional Administrator: Charles
    Meyrick

U.S.
Department of Transportation
Federal Railroad Administration
Region III
536 South Clark Street
Room 210
Chicago, IL   60605
  (312) 353-6203
    Regional Administrator: Jacob Sharp

U.S.
Department of Transportation
Federal Railroad Administration
Region IV
Federal Office Building
Room 11A23
819 Taylor Street
Fort Worth, TX   76102
  (817) 334-3601
    Regional Administrator: Trinidad
    Gullen

U.S.
Department of Transportation
Federal Railroad Administration
Region V
2 Embarcadero Center
Suite 630
San Francisco, CA   94111
  (415) 556-6411

Regional Administrator: Earl
  Anderson

U.S.
Department of Transportation
Federal Railroad Administration
Research and Development Office
2100 Second Street, S.W.
Washington, DC   20590
  (202) 426-0955
    Technology Planning Officer: Ned
    Ahmed

U.S.
Department of Transportation
Materials Transportation Bureau
2100 Second Street, S.W.
Washington, DC   20590
  (202) 755-9260
  Director

U.S.
Department of Transportation
National Highway Traffic Safety Ad-
  ministration
Technical Reference Branch
400 Seventh Street, S.W.
Room 5108
Washington, DC   20590
  (202) 426-2768
  Winfred Desmond

U.S.
Department of Transportation
Office of Environmental Affairs
400 Seventh Street, S.W.
Washington, DC   20590
  (202) 426-4357
  Director: Martin Convisser

U.S.
Department of Transportation
Office of the Secretary
Environment and Safety Office
400 Seventh Street, S.W.
Washington, DC   20590
  (202) 426-4492
    Transportation Specialist: George
    McDonald

U.S.
Department of Transportation

Office of the Secretary
Research and Development Plans
400 Seventh Street, S.W.
Washington, DC 20590
  (202) 426-0975
  Director

U.S.
Department of Transportation
Regional Office I
Transportation Systems Center
55 Broadway
Cambridge, MA 02142
  (617) 494-2709
  Regional Representative of the
    Secretary

U.S.
Department of Transportation
Regional Office II
26 Federal Plaza
New York, NY 10007
  (212) 264-2672
  Regional Representative of the
    Secretary

U.S.
Department of Transportation
Regional Office III
434 Walnut Street
Philadelphia, PA 19106
  (215) 597-9430
  Regional Representative of the
    Secretary

U.S.
Department of Transportation
Regional Office IV
1720 Peachtree Road, N.W.
Atlanta, GA 30309
  (404) 881-2738
  Regional Representative of the
    Secretary

U.S.
Department of Transportation
Regional Office V
300 South Wacker Drive
Chicago, IL 60606
  (312) 353-4000
  Regional Representative of the
    Secretary

U.S.
Department of Transportation
Regional Office VI
9-C-18 Federal Center
1100 Commerce Street
Dallas, TX 75242
  (214) 749-1851
  Regional Representative of the
    Secretary

U.S.
Department of Transportation
Regional Office VII
601 East 12th Street
Kansas City, MO 64106
  (816) 374-5801
  Regional Representative of the
    Secretary

U.S.
Department of Transportation
Regional Office VIII
Prudential Plaza
1050 17th Street
Denver, CO 80202
  (303) 837-3242
  Regional Representative of the
    Secretary

U.S.
Department of Transportation
Regional Office IX
2 Embarcadero Center
San Francisco, CA 94111
  (415) 556-5961
  Regional Representative of the
    Secretary

U.S.
Department of Transportation
Regional Office X
3112 Federal Building
915 Second Avenue
Seattle, WA 98174
  (206) 442-0590
  Regional Representative of the
    Secretary

U.S.
Department of Transportation
Research and Special Programs Admin-
  istration

Transportation System Center
400 Seventh Street, S.W.
Washington, DC 20590
  (202) 426-4000
  Director: James Costantino

U.S.
Department of Transportation
Transportation Programs Bureau
400 Seventh Street, S.W.
Washington, DC 20590
  (202) 426-1640
  Director

U.S.
Department of Transportation
Transportation Systems Center
Information Division
Kendall Square
Cambridge, MA 02142
  (617) 494-2654
  Chief: Robert Tap

U.S.
Department of Transportation
Transportation Systems Center
Technology Sharing Office
Kendall Square
Cambridge, MA 02142
  (617) 494-2486
  Director

U.S.
Department of Transportation
Urban Mass Transportation Admin-
  istration
400 Seventh Street, S.W.
Washington, DC 20590
  (202) 426-4043
  Administrator: Richard Page

U.S.
Department of Transportation
Urban Mass Transportation Admin-
  istration
Human Resources and Technical
  Division
2100 Second Street, S.W.
Washington, DC 20590
  (202) 426-9274
  Program Manager: Dr. Frank Enty

U.S.
Department of Transportation
Urban Mass Transportation Admin-
  istration
Planning, Management, and Demonstra-
  tion
400 Seventh Street, S.W.
Washington, DC 20590
  (202) 426-4000
  Associate Administrator: Robert
  McManus

U.S.
Department of Transportation
Urban Mass Transportation Admin-
  istration
Region I
c/o Transportation System Center
53 Broadway
Cambridge, MA 02142
  (617) 494-2055
  Director

U.S.
Department of Transportation
Urban Mass Transportation Admin-
  istration
Region II
26 Federal Plaza
New York, NY 10007
  (212) 264-8162
  Director

U.S.
Department of Transportation
Urban Mass Transportation Admin-
  istration
Region III
434 Walnut Street
Philadelphia, PA 19106
  (215) 597-8098
  Director

U.S.
Department of Transportation
Urban Mass Transportation Admin-
  istration
Region IV
1720 Peachtree Street, N.W.
Atlanta, GA 30309
  (404) 881-3948
  Director

U.S.
Department of Transportation
Urban Mass Transportation Admin-
  istration
Region V
300 South Wacker Drive
Chicago, IL 60606
  (312) 353-2789
  Director

U.S.
Department of Transportation
Urban Mass Transportation Admin-
  istration
Region VI
819 Taylor Street
Fort Worth, TX 76102
  (817) 334-3787
  Director

U.S.
Department of Transportation
Urban Mass Transportation Admin-
  istration
Region VII
6301 Rock Hill Road
Kansas City, MO 64131
  (816) 926-5053
  Director

U.S.
Department of Transportation
Urban Mass Transportation Admin-
  istration
Region VIII
1050-17th Street
Denver, CO 80202
  (303) 837-3242
  Director

U.S.
Department of Transportation
Urban Mass Transportation Admin-
  istration
Region IX
2 Embarcadero Center
San Francisco, CA 94111
  (415) 556-2884
  Director

U.S.
Department of Transportation

Urban Mass Transportation Administration
Region X
915 Second Avenue
Seattle, WA 98174
  (206) 442-4210
  Director

U.S.
Department of Transportation
Urban Mass Transportation Admin-
  istration
Technology Development and Deployment
400 Seventh Street, S.W.
Washington, DC 20590
  (202) 426-4000
  Associate Administrator: George
    Pastor

U.S.
Interstate Commerce Commission
Energy and Environment Branch
Environmental Transportation
Washington, DC 20423
  (202) 275-7658
  Specialist: David Rector

Utah
Environmental Coordinating Committee
Department of Transportation
124 State Capitol
Salt Lake City, UT 84111
  (801) 533-5794
  Gene Sturzenegger

Virginia
Department of Highways and Transpor-
  tation
1221 East Broad Street
Room 1114
Richmond, VA 23219
  (804) 786-4304
  Environmental Quality Engineer:
    R.L. Hundley

Washington
Transportation Department
Highway Administration Building
Olympia, WA 98501
  (206) 753-6005
  Director

<u>Wisconsin</u>
Transportation Department
1208 Hill Farms
State Office Building
4802 Sheboygan Avenue
Madison, WI 53702
  (608) 266-1113
  Director

APPENDIX

APPENDIX A: EIS-RELATED LIBRARIES

Alameda Free Library
2264 Santa Clara Avenue
Alameda, CA 94501
   (415) 522-5413
   Head Librarian

Alaska
Department of Environmental Conser-
   vation
Library
Pouch O
Juneau, AK 99811
   (907) 465-2606
   Librarian: Sheila Zagars

Albany Public Library
161 Washington Avenue
Albany, NY 12210
   (518) 449-3380
   Head Librarian

Alexandria Library
717 Queen Street
Alexandria, VA 23314
   (703) 750-6351
   Head Librarian

Alhambra Public Library
Main Street
Alhambra, CA 91801
   (213) 289-4216
   Head Librarian

American Library Association
50 East Huron Street

Chicago, IL 60611
   (312) 944-6780
   Joel Lee

Anaheim Public Library
500 West Broadway
Anaheim, CA 92805
   (714) 533-5221
   Head Librarian

Anderson Public Library
32 West Tenth Street
Anderson, IN 46016
   (317) 644-0938
   Head Librarian

Appleton Public Library
121 South Oneida Street
Appleton, WI 54911
   (414) 734-7171
   Head Librarian

Arlington County Department of
Libraries
1015 North Quincy Street
Arlington, VA 22201
   (703) 527-4777
   Head Librarian

Berkeley Public Library
2090 Kittredge Street
Berkeley, CA 94704
   (415) 644-6095
   Head Librarian

Berkshire Athenaeum
One Wendell Avenue
Pittsfield, MA   01201
   (413) 442-1559
   Head Librarian

Binghamton Public Library
78 Exchange Street
Binghamton, NY   13901
   (607) 723-6457
   Head Librarian

Boston Public Library and Eastern
   Massachusetts Regional Public
   Library System
666 Boylston Street
Boston, MA   02117
   (617) 536-5400
   Head Librarian

Bridgeport Public Library
925 Broad Street
Bridgeport, CT   06603
   (203) 576-7777
   Head Librarian

Bristol Public Library
5 High Street
Bristol, CT   06010
   (203) 582-9505
   Head Librarian

Brockton Public Library
304 Main Street
Brockton, MA   02401
   (617) 587-2515
   Head Librarian

Brooklyn Public Library
Grand Army Plaza
Brooklyn, NY   11238
   (212) 636-3111
   Head Librarian

Brown County Library
515 Pine Street
Green Bay, WI   54301
   (414) 497-3450
   Head Librarian

Buena Park Library District
7150 La Palma Avenue

Buena Park, CA   90620
   (714) 826-4100
   Head Librarian

Buffalo and Erie County Public Library
Lafayette Square
Buffalo, NY   14203
   (716) 856-7525
   Head Librarian

Burbank Public Library
110 North Glenoaks Boulevard
Burbank, CA   91503
   (213) 847-9737
   Head Librarian

California Academy of Sciences
   Library
Golden Gate Park
San Francisco, CA   94118
   (415) 221-5100
   Head Librarian

California
Department of Fish and Game
Marine Technical Information Center
350 Golden Shore
Long Beach, CA   90802
   (213) 435-7741
   Director

California
State Library
Library and Courts Building
P.O. Box 2037
Sacramento, CA   95809
   (916) 445-2585
   Head Librarian

California
State Resources Agency Library
1416 Ninth Street
Room 117
Sacramento, CA   95814
   (916) 445-7752
   Head Librarian

California State University Library
Central University Library
San Diego, CA   92182
   (714) 452-3061
   Head Librarian

Cambridge Public Library
449 Broadway
Cambridge, MA   02138
  (617) 876-5005
  Head Librarian

Charlotte and Mecklenburg County
  Public Library
310 North Tryon Street
Charlotte, NC   28202
  (704) 374-2725
  Head Librarian

Chesapeake Public Library System
300 Cedar Road
Chesapeake, VA   23320
  (804) 547-6579
  Head Librarian

Chicopee Public Library
Market Square
Chicopee, MA   01013
  (413) 594-6679
  Head Librarian

Chula Vista Public Library
365 F Street
Chula Vista, CA   92010
  (714) 575-5062
  Head Librarian

Connecticut
State Library
231 Capitol Avenue
Hartford, CT   06106
  (203) 566-4777
  Head Librarian

Cumberland County Public Library
215 Anderson Street
Fayetteville, NC   28302
  (919) 483-8600
  Head Librarian

Daly City Public Library
40 Wembley Drive
Daly City, CA   94015
  (415) 992-4500
  Head Librarian

Danbury Public Library
170 Main Street

Box 1111
Danbury, CT   06810
  (203) 792-0260
  Head Librarian

Davis Conservation Library
404 East Main
P.O. Box 776
League City, TX   77573
  (713) 332-3402
  Librarian: Marion Yang Kwok

Denver Public Library
Conservation Library
1357 Broadway
Denver, CO   80203
  (303) 573-5152
  Conservation Specialist: Kay
  Collins

Denver Public Library
Fish and Wildlife Reference Service
2100 West Mississippi Avenue
Denver, CO   80223
  (303) 922-0505
  Marsha Fralick

Downey City Library
8490 Third Street
Downey, CA   90241
  (213) 923-3256
  Head Librarian

Duluth Public Library
101 West Second Street
Duluth, MN   55802
  (218) 722-5803
  Head Librarian

Durham City-County Public Library
311 East Main Street
Durham, NC   27702
  (919) 682-9109
  Head Librarian

East Hartford Public Library
840 Main Street
East Hartford, CT   06108
  (203) 289-6429
  Head Librarian

Ecology Forum, Inc.
Environmental Information Center
Suite 303 East
200 Park Avenue
New York, NY 10017
  President: James Kolleggar

Enoch Pratt Free Library
400 Cathedral Street
Baltimore, MD 21201
  (301) 396-5430
  Head Librarian

Environmental Law Institute
Library
1346 Connecticut Avenue, N.W.
Washington, DC 20036
  (202) 452-9600
  Librarian: Barbara Rodes

Evansville Public Library
22 S.E. Fifth Street
Evansville, IN 47708
  (812) 425-2621
  Head Librarian

Everett Public Library
2702 Hoyt Avenue
Everett, WA 98201
  (206) 259-8857
  Head Librarian

Fairfield Public Library
1080 Old Post Road
Fairfield, CT 06430
  (203) 259-8303
  Head Librarian

Fall River Public Library
104 North Main Street
Fall River, MA 02720
  (617) 676-8541
  Head Librarian

Forsyth County Public Library System
660 West Fifth Street
Winston-Salem, NC 27101
  (919) 727-2556
  Head Librarian

Fort Wayne and Allen County Public
  Library

900 Webster Street
Fort Wayne, IN 46802
  (219) 424-7241
  Head Librarian

Framingham Public Library
Worcester Road
Framingham, MA 01701
  (617) 872-7432
  Head Librarian

Fresno County Free Library
2420 Mariposa Street
Fresno, CA 93721
  (209) 488-3191
  Head Librarian

Fullerton Public Library
Commonwealth Avenue
Fullerton, CA 92632
  (714) 871-9440
  Head Librarian

Gary Public Library
220 West Fifth Avenue
Gary, IN 46402
  (219) 886-2484
  Head Librarian

Gilbert Simmons Public Library
711-59th Place
Kenosha, WI 53140
  (414) 657-6101
  Head Librarian

Glendale Public Library
222 Harvard Street
Glendale, CA 91205
  (213) 965-2020
  Head Librarian

Great Falls Public Library
Third Street and Second Avenue North
Great Falls, MT 59401
  (406) 453-0349
  Head Librarian

Greensboro Public Library
201 Greene Street
Greensboro, NC 27402
  (919) 373-2471
  Head Librarian

Greenwich Library
101 West Putnam Avenue
Greenwich, CT 06830
   (203) 622-7900
   Head Librarian

Hammond Public Library
566 State Street
Hammond, IN 46320
   (219) 931-5100
   Head Librarian

Hartford Public Library
500 Main Street
Hartford, CT 06103
   (203) 525-9121
   Head Librarian

Hawaii
State Library System
478 South King Street
Honolulu, HI 96813
   (808) 548-4775
   Head Librarian

Hayward Public Library
22734 Mission Boulevard
Hayward, CA 94541
   (415) 581-5464
   Head Librarian

High Point Public Library
411 South Main Street
High Point, NC 27262
   (919) 882-9225
   Head Librarian

Holyoke Public Library
335 Maple Street
Holyoke, MA 01040
   (413) 534-3357
   Head Librarian

Huntington Beach Public Library
7111 Talbert
Huntington Beach, CA 92648
   (714) 842-4481
   Head Librarian

Indiana
State Library
Library and Historical Building

140 North Senate
Indianapolis, IN 46204
   (317) 633-5441
   Head Librarian

Indianapolis-Marion County Public
   Library
40 E Street
Indianapolis, IN 46204
   (317) 635-5662
   Head Librarian

Indiana University
School of Public Environmental
   Affairs Library
400 East Seventh Street
Bloomington, IN 47401
   (812) 337-8552
   Head Librarian

Inglewood Public Library
Manchester Street
Inglewood, CA 90301
   (213) 649-7380
   Head Librarian

Jacob Oliphant Library
Air Pollution Division
State Board of Health Building
1330 West Michigan Street
Indianapolis, IN 46206
   (317) 633-4610
   Head Librarian

Jervis Library Association
613 North Washington Street
Rome, NY 13440
   (315) 336-4570
   Head Librarian

Kent State University
Center for Urban Regionalism
Library and Information Services
Kent, OH 44240
   Librarian

Kern County Library
1315 Truxtun Avenue
Bakersfield, CA 93301
   (805) 861-2130
   Head Librarian

La Crosse County Library
400 North Fourth Street
La Crosse County, WI   54601
  (608) 785-9638
  Head Librarian

Lawrence Free Public Library
51 Lawrence Street
Lawrence, MA   01841
  (617) 682-1727
  Head Librarian

Long Beach Public Library
101 Pacific Avenue
Long Beach, CA   90802
  (213) 437-2949
  Head Librarian

Los Angeles Public Library
630 West Fifth Street
Los Angeles, CA   90017
  (213) 626-7461
  Head Librarian

Lowell City Library
401 Merrimack Street
Lowell, MA   01852
  (617) 454-8821
  Head Librarian

Lynchburg Public Library
914 Main Street
Lynchburg, VA   24504
  (804) 847-1565
  Head Librarian

Lynn Public Library
5 North Common Street
Lynn, MA   01902
  (617) 595-0567
  Head Librarian

Madison Public Library
201 West Mifflin Street
Madison, WI   53703
  (608) 266-6300
  Head Librarian

Malden Public Library
36 Salem Street
Malden, MA   02148
  (617) 324-0218

  Head Librarian

Maryland
Department of Health and Mental
  Hygiene
Library
201 West Preston Street
Baltimore, MD   20201
  (301) 383-2634
  Head Librarian

Maryland
Department of Natural Resources
Library
Tawes State Office Building
Annapolis, MD   21401
  (301) 267-5015
  Head Librarian

Maryland
State Library
Court of Appeals Building
Annapolis, MD   21401
  (301) 269-3395
  Head Librarian

Massachusetts
State Library
George Fingold Library
State House
Boston, MA   02133
  (617) 727-2590
  Head Librarian

Medford Public Library
111 High Street
Medford, MA   02155
  (617) 395-7950
  Head Librarian

Meridan Public Library
105 Miller Street
Meridan, CT   06450
  (203) 238-2344
  Head Librarian

Midwest Research Institute
Library
425 Volker Boulevard
Kansas City, MO   64110
  (816) 753-7600
  Head Librarian

Milford Public Library
Taylor Library Building
5 Broad Street
Milford, CT   06460
  (203) 878-7461
  Head Librarian

Milwaukee Public Library
814 West Wisconsin Avenue
Milwaukee, WI   53233
  (414) 278-3000
  Head Librarian

Minneapolis Public Library
300 Nicollet Mall
Minneapolis, MN   55401
  (612) 372-6500
  Head Librarian

Montana
State Library
930 East Lyndale
Helena, MT   59601
  (406) 449-3004
  Head Librarian

Moss Landing Marine Laboratories
Library
P.O. Box 223
Moss Landing, CA   95039
  (408) 633-3304
  Head Librarian

Mountain View Public Library
585 Franklin Street
Mountain View, CA   94040
  (415) 968-6595
  Head Librarian

Mount Vernon Public Library
28 South First Street
Mount Vernon, NY   10550
  (914) 688-1840
  Head Librarian

Muncie Public Library
301 East Jackson Street
Muncie, IN   47305
  (317) 288-9971
  Head Librarian

National League of Cities

Library
1620 Eye Street, N.W.
Washington, DC   20006
  (202) 293-7177
  Reference Librarian

New Bedford Free Public Library
Pleasant Street
New Bedford, MA   02741
  (617) 999-6291
  Head Librarian

New Britain Public Library
West Main and High Streets
New Britain, CT   06050
  (203) 224-3155
  Head Librarian

New Haven Free Public Library
133 Elm Street
New Haven, CT   06510
  (203) 562-0151
  Head Librarian

Newport News Public Library
110 Main Street
Newport News, VA   23601
  (804) 596-5723
  Head Librarian

New Rochelle Public Library
662 Main Street
New Rochelle, NY   10805
  (914) 632-7878
  Head Librarian

New York Botanical Garden
Library
200th Street and Southern Boulevard
Bronx, NY   10458
  (212) 220-8751
  Charles Long

New York Public Library
Fifth Avenue and 42nd Street
New York, NY   10018
  (212) 790-6161
  Head Librarian

Newton Free Library
414 Centre Street
Newton, MA   01201

(617) 527-7700
Head Librarian

Niagara Falls Public Library
1425 Main
Niagara Falls, NY 14301
  (716) 278-8092
  Head Librarian

Norfolk Public Library
301 East City Hall Avenue
Norfolk, VA 23510
  (804) 441-2429
  Head Librarian

Norwalk Public Library
1st District
One Belden Avenue
Norwalk, CT 06850
  (203) 866-5559
  Head Librarian

Oakland Public Library
125-14th Street
Oakland, CA 94512
  (415) 273-3134
  Head Librarian

Onondaga County Public Library
335 Montgomery Street
Syracuse, NY 13202
  (315) 473-4489
  Head Librarian

Ontario Public Library
215 East C Street
Ontario, CA 91761
  (714) 984-2758
  Head Librarian

Orange County Public Library
431 City Drive
Orange, CA 92668
  (714) 634-7841
  Head Librarian

Oshkosh Public Library
106 Washington Avenue
Oshkosh, WI 54901
  (414) 424-0473
  Head Librarian

Oxnard Public Library
214 South C Street
Oxnard, CA 93030
  (805) 487-3981
  Head Librarian

Pack Memorial Public Library
Pack Square
Asheville, NC 28801
  (704) 252-8701
  Head Librarian

Palo Alto City Library
1213 Newell Road
Palo Alto, CA 94303
  (415) 329-2436
  Head Librarian

Parmly Billings Public Library
510 North Broadway
Billings, MT 59101
  (406) 248-7391
  Head Librarian

Pasadena Public Library
285 East Walnut Street
Pasadena, CA 91101
  (213) 577-4066
  Head Librarian

Pomona Public Library
625 South Garey Avenue
Pomona, CA 91766
  (714) 620-2033
  Head Librarian

Portsmouth Public Library
601 Court Street
Portsmouth, VA 23704
  (804) 393-8501
  Head Librarian

Queensborough Public Library
89-11 Merick Boulevard
Jamaica, NY 11432
  (212) 739-1900
  Head Librarian

Racine Public Library
75 Seventh Street
Racine, WI 53403
  (414) 636-9241

Head Librarian

Redondo Beach Public Library
309 Esplanade
Redondo Beach, CA   90277
    (213) 376-8723
    Head Librarian

Redwood City Public Library
881 Jefferson Avenue
Redwood City, CA   94063
    (415) 369-3738
    Head Librarian

Richmond Public Library
Civic Center Plaza
Richmond, CA   94804
    (415) 234-6632
    Head Librarian

Richmond Public Library
101 East Franklin Street
Richmond, VA   23219
    (804) 780-4256
    Head Librarian

Riverside City and County Public
    Library
3851 Seventh Street
Riverside, CA   92502
    (714) 787-7201
    Head Librarian

Roanoke Public Library
706 South Jefferson Street
Roanoke, VA   24011
    (703) 981-2473
    Head Librarian

Robbins Library
700 Massachusetts Avenue
Arlington, MA   02174
    (617) 643-0026
    Head Librarian

Rochester Public Library
Broadway at First Street, S.E.
Rochester, MN   55901
    (507) 288-9070
    Head Librarian

Saint Paul Public Library

90 West Fourth Street
Saint Paul, MN   55102
    (612) 224-3383
    Head Librarian

Salinas Public Library
110 West San Luis Street
Salinas, CA   93901
    (408) 758-7311
    Head Librarian

San Bernardino Public Library
401 North Arrowhead Avenue
San Bernardino, CA   92401
    (714) 889-0264
    Head Librarian

San Diego Public Library
820 E Street
San Diego, CA   92101
    (714) 236-5800
    Head Librarian

San Francisco Public Library
Civic Center
San Francisco, CA   94102
    (415) 558-3191
    Head Librarian

San Jose Public Library
180 West San Carlos Street
San Jose, CA   95113
    (408) 277-4000
    Head Librarian

San Leandro Community Library Center
300 Estudillo Avenue
San Leandro, CA   94577
    (415) 483-1511
    Head Librarian

San Mateo Public Library
55 West Third Avenue
San Mateo, CA   94402
    (415) 574-6950
    Head Librarian

Santa Ana Public Library
502 Civic Center Drive, West
Santa Ana, CA   92701
    (714) 834-4013
    Head Librarian

Santa Barbara Public Library
40 East Anapamu Street
Box 1019 Z
Santa Barbara, CA   93102
   (805) 962-7653
   Head Librarian

Santa Clara City Library
2635 Homestead Road
Santa Clara, CA   95051
   (408) 984-3097
   Head Librarian

Santa Monica Public Library
1343 Sixth Street
Santa Monica, CA   90401
   (213) 451-5751
   Head Librarian

Schenectady County Public Library
Liberty and Clinton Streets
Schenectady, NY   12305
   (518) 382-3542
   Head Librarian

Seattle Public Library
1000 Fourth Avenue
Seattle, WA   98104
   (206) 625-2665
   Head Librarian

Silas Bronson Public Library
267 Grand Street
Waterbury, CT   06702
   (203) 574-8200
   Head Librarian

Sioux Falls Public Library
201 North Main
Sioux Falls, SD   57101
   (605) 339-7081
   Head Librarian

Somerville Public Library
Highland Avenue and Walnut Street
Somerville, MA   02143
   (617) 623-5000
   Head Librarian

Sonoma County Library
Third and E Streets
Santa Rosa, CA   95404

   (707) 545-0831
   Head Librarian

South Bend Public Library
122 West Wayne Street
South Bend, IN   46601
   (219) 288-4413
   Head Librarian

Southern California Air Pollution
   Control District Library
434 South San Pedro Street
Los Angeles, CA   90013
   (213) 974-7426
   Head Librarian

Southwestern Illinois Metropolitan
   and Regional Planning Commission
Library
203 West Main Street
Collinsville, IL   62234
   (618) 344-4250
   Librarian: Judy Thompson

Spokane Public Library
906 Main Avenue
Spokane, WA   99201
   (509) 838-3361
   Head Librarian

Springfield City Library
220 State Street
Springfield, MA   01103
   (413) 739-3871
   Head Librarian

Stanford Public Library
96 Broad Street
Stanford, CT   06901
   (203) 325-4354
   Head Librarian

Stanford University
Hopkins Marine Station Library
Pacific Grove, CA   93950
   (408) 373-0464
   Head Librarian

Stanislaus County Free Public Library
1402 Eye Street
McHenry Library Building
Modesto, CA   95354

(209) 526-6822
Head Librarian

Stockton and San Joaquin County
  Public Library
605 North El Dorado
Stockton, CA 95202
  (209) 944-8415
  Head Librarian

Stratford Library Association
2203 Main Street
Stratford, CT 06497
  (203) 378-7345
  Head Librarian

Sunnyvale Public Library
665 West Olive
Box 670 Z
Sunnyvale, CA 94086
  (408) 245-9171
  Head Librarian

Taylor Memorial Library
4205 Victoria Boulevard
Hampton, VA 23669
  (804) 727-6234
  Head Librarian

Tennessee Valley Authority
Library
400 Commerce Avenue
Knoxville, TN 37902
  (615) 632-3464
  Librarian: Jesse Mills

Thomas Crane Public Library
40 Washington Street
Quincy, MA 02169
  (617) 471-2400
  Head Librarian

Torrance Public Library
3031 Torrance Boulevard
Torrance, CA 90503
  (213) 328-2251
  Head Librarian

Troy Public Library
100 Second
Troy, NY 12180
  (518) 274-7071

Head Librarian

Tufts Library
46 Broad Street
Weymouth, MA 02188
  (617) 337-1402
  Head Librarian

United States Conference of Mayors
Library
1620 Eye Street, N.W.
Washington, DC 20006
  (202) 293-7177
  Head Librarian

U.S.
Department of Agriculture
National Agricultural Library
Beltsville, MD 20705
  (301) 344-3778
  Director of Library Services

U.S.
Department of HEW
National Library of Medicine
9000 Rockville Pike
Bethesda, MD 20205
  (301) 495-4000
  Director: Martin Cummings

U.S.
Department of the Interior
Bureau of Reclamation
Information Storage and Retrival Pro-
  ject
Building 67
Denver Federal Center
Denver, CO 80225
  (303) 234-4441
  Director

U.S.
Department of the Interior
Geological Survey Library
National Center
Mail Stop 950
Reston, VA 22092
  (703) 860-6671
  Head Librarian

U.S.
Department of the Interior

Geological Survey
USGS Library
Box 25046
Denver Federal Center
Stop 914
Denver, CO 80225
  (303) 234-4133
  Librarian: Irvil Shultz

U.S.
Department of the Interior
Geological Survey
Western Regional Library
345 Middlefield Road
Menlo Park, CA 94025
  (415) 323-8111
  Librarian

U.S.
Department of the Interior
Office of Library and Information
  Service
C Street Between 18th and 19th
  Streets, N.W.
Washington, DC 20240
  (202) 343-1100
  Director: Mary Huffer

U.S.
Library of Congress
Library Environment Resources
  Office
10 First Street, S.E.
Washington, DC 20540
  (202) 287-5000
  Director: Paul Berry

U.S.
Library of Congress
National Referral Center
Research Services
Washington, DC 20540
  (202) 287-5670
  Director

U.S.
Library of Congress
National Referral Center
Science and Technology Division
10 First Street, S.E.
Washington, DC 20540
  (202) 426-5670

Director

U.S.
National Commission on Libraries and
  Information Sciences
Suite 601
1717 K Street, N.W.
Washington, DC 20036
  (202) 653-6252
  Director

University of California
Berkeley
Natural Resources Libraries
40 Giannini Hall
Berkeley, CA 94720
  (415) 642-4493
  Head Librarian

University of California
Davis
Department of Environmental Toxicol-
  ogy
Environmental Toxicology Library
Davis, CA 95616
  (916) 752-2562
  Head Librarian

University of California
Santa Cruz
Environmental Studies Library
Room 317
Social Sciences Building
Santa Cruz, CA 95060
  (408) 429-2104
  Head Librarian

University of Montana
Environmental Library
Room 208A
Natural Sciences Building
Missoula, MT 59801
  (406) 243-2282
  Head Librarian

Utica Public Library
303 Genessee Street
Utica, NY 13501
  (315) 735-2279
  Head Librarian

Vigo County Public Library

222 North Seventh Street
Terre Haute, IN 47801
 (812) 232-5041
 Head Librarian

Virginia
State Library
11th and Capitol Street
Richmond, VA 23219
 (804) 786-8920
 Head Librarian

Wake County Public Library
104 Fayetteville Street
Raleigh, NC 27601
 (919) 755-6077
 Head Librarian

Waltham Public Library
735 Main Street
Waltham, MA 02154
 (617) 893-1750
 Head Librarian

Washington
State Library
State Library Building
Olympia, WA 98504
 (206) 753-5590
 Head Librarian

Wauwatosa Public Library
7635 West North Avenue
Wauwatosa, WI 53213
 (414) 258-5700
 Head Librarian

West Allis Public Library
1508 South 75th Street
West Allis, WI 53214
 (414) 476-6550
 Head Librarian

Westchester Library System
285 Central Avenue
Hartsdale, NY 10606
 (914) 761-7620
 Head Librarian

West Hartford Public Library
20 South Main Street
West Hartford, CT 06407

 (203) 236-4561
 Head Librarian

West Haven Public Library
300 Elm Street
West Haven, CT 06511
 (203) 932-2221
 Head Librarian

Whittier Public Library
7344 South Washington Avenue
Whittier, CA 90602
 (213) 698-8181
 Head Librarian

Worcester Public Library and Central
 Massachusetts Regional Library
 System Headquarters
Salem Square
Worcester, MA 01608
 (617) 752-3751
 Head Librarian

Yale University School of Forestry
Library
205 Prospect Street
New Haven, CT 06511
 (203) 436-0577
 Head Librarian

Yonkers Public Library
Broadway Branch
70 South Broadway
Yonkers, NY 10701
 (914) 337-1500
 Head Librarian

APPENDIX B: EIS-RELATED NEWSPAPERS

Albany Knickerbocker News-Union Star
645 Albany-Shaker Road
Albany, NY 12201
   (518) 454-5694
   Environmental Editor

Albany Times-Union
645 Albany-Shaker Road
Albany, NY 12201
   (518) 454-5694
   Environmental Editor

Bakersfield Californian
1707 Eye Street
Box 440
Bakersfield, CA 93302
   (805) 395-7330
   Environmental Editor

Baltimore News-American
Lombard and South Streets
Baltimore, MD 21202
   (301) 752-1212
   Environmental Editor

Baltimore Sun
Calvert and Centre Streets
Baltimore, MD 21203
   (301) 332-6000
   Environmental Editor

Billings Gazette
401 North Broadway
Billings, MT 59103
   (406) 245-3071

Environmental Editor

Binghamton Press
Vestal Parkway East
Binghamton, NY 13902
   (607) 798-1234
   Environmental Editor

Boston Christian Science Monitor
One Norway Street
Boston, MA 02115
   (617) 262-2300
   Environmental Editor

Boston Globe
135 Morrissey Boulevard
Boston, MA 02107
   (617) 929-2000
   Environmental Editor

Boston Herald American
300 Harrison Avenue
Boston, MA 02106
   (617) 426-3000
   Environmental Editor

Bridgeport Post
410 State Street
Bridgeport, CT 06602
   (203) 333-0161
   Environmental Editor

Brockton Enterprise and Times
60 Main Street
Brockton, MA 02403

    (617) 586-6200
    Environmental Editor

Buffalo Courier-Express
785 Main Street
Buffalo, NY 14240
    (716) 847-5353
    Environmental Editor

Buffalo News
One News Plaza
Buffalo, NY 14240
    (716) 849-3434
    Environmental Editor

Charlotte News
600 South Tyron Street
Box 2138
Charlotte, NC 28201
    (704) 374-7471
    Environmental Editor

Charlotte Observer
600 South Tryon Street
Box 2188
Charlotte, NC 28201
    (704) 374-7471
    Environmental Editor

Duluth News Tribune
424 West First Street
Duluth, MN 55801
    (218) 723-5220
    Environmental Editor

Evansville Courier
201 N.W. Second Street
Box 268
Evansville, IN 47701
    (812) 424-7711
    Environmental Editor

Everett Herald
Graud and California Streets
P.O. Box 930
Everett, WA 98206
    (206) 259-5151
    Environmental Editor

Fort Wayne Journal-Gazette
600 West Main Street
Fort Wayne, IN 46802

    (219) 461-8333
    Environmental Editor

Fort Wayne News-Sentinel
600 West Main Street
Fort Wayne, IN 46802
    (219) 461-8333
    Environmental Editor

Fresno Bee
1626 E Street
Fresno, CA 93786
    (209) 268-5221
    Environmental Editor

Garden City Newsday
550 Stewart Avenue
Garden City, NY 11530
    (516) 222-5388
    Environmental Editor

Gary Post-Tribune
1065 Broadway
Gary, IN 46402
    (219) 886-5157
    Environmental Editor

Green Bay Press-Gazette
Box 430
Green Bay, WI 54305
    (414) 435-4411
    Environmental Editor

Greensboro News
200 East Market Street
P.O. Box 20848
Greensboro, NC 27420
    (919) 373-1000
    Environmental Editor

Hammond Times
1065 Broadway
Hammond, IN 46402
    (219) 886-5157
    Environmental Editor

Hartford Courant
285 Broad Street
Hartford, CT 06115
    (203) 249-6411
    Environmental Editor

Honolulu Advertiser
News Building
605 Kapiolani Boulevard
Box 3350
Honolulu, HI 96801
  (808) 525-8000
  Environmental Editor

Honolulu Star-Bulletin
News Building
605 Kapiolani Boulevard
Box 3350
Honolulu, HI 96801
  (808) 525-8000
  Environmental Editor

Indi napolis News
307 rth Pennsylvania Street
Indianapolis, IN 46206
  (317) 633-1154
  Environmental Editor

Indianapolis Star
307 North Pennsylvania Street
Indianapolis, IN 46206
  (317) 633-1154
  Environmental Editor

Lawrence Eagle-Tribune
P.O. Box 100
Lawrence, MA 01842
  (617) 685-1000
  Environmental Editor

Long Beach Independent
604 Pine Avenue
Box 230
Long Beach, CA 90844
  (213) 435-1161
  Environmental Editor

Long Beach Press-Telegram
604 Pine Avenue
Box 230
Long Beach, CA 90844
  (213) 435-1161
  Environmental Editor

Los Angeles Herald-Examiner
1111 South Broadway
Box 2416 Terminal Annex
Los Angeles, CA 90051

  (213) 748-1212
  Environmental Editor

Los Angeles Times
Times-Mirror Square
Los Angeles, CA 90053
  (213) 972-5000
  Environmental Editor

Lowell Sun
15 Kearney Square
Lowell, MA 01852
  (617) 455-5671
  Environmental Editor

Middletown Times Herald-Record
40 Mulberry Street
Middletown, NY 10940
  (914) 343-2181
  Environmental Editor

Milwaukee Journal
333 West State Street
Milwaukee, WI 53201
  (414) 224-2000
  Environmental Editor

Milwaukee Sentinel
333 West State Street
Milwaukee, WI 53201
  (414) 224-2000
  Environmental Editor

Minneapolis Star
427 Portland Avenue
Minneapolis, MN 55488
  (612) 372-4141
  Environmental Editor

Minneapolis Tribune
427 Portland Avenue
Minneapolis, MN 55488
  (612) 372-4141
  Environmental Editor

Modesto Bee
14th and H Streets
Box 3928
Modesto, CA 95352
  (209) 524-4041
  Environmental Editor

New Haven Register
367 Orange Street
New Haven, CT 06503
   (203) 562-1121
   Environmental Editor

Newport News Press
7505 Warwick Boulevard
Box 746
Newport News, VA 23607
   (804) 244-8424
   Environmental Editor

New York Daily News
220 East 42nd Street
New York, NY 10017
   (212) 949-1234
   Environmental Editor

New York Post
210 South Street
New York, NY 10002
   (212) 349-5000
   Environmental Editor

New York Times
229 West 43rd Street
New York, NY 10036
   (212) 556-1234
   Environmental Editor

New York Wall Street Journal
22 Cortlandt Street
New York, NY 10007
   (212) 285-5000
   Environmental Editor

Norfolk Ledger-Star
150 West Brambleton Avenue
Norfolk, VA 23501
   (804) 446-2000
   Environmental Editor

Norfolk Virginian-Pilot
150 West Brambleton Avenue
Norfolk, VA 23501
   (804) 446-2000
   Environmental Editor

Oakland Tribune
409-13th Street
Box 24304

Oakland, CA 94623
   (415) 645-2000
   Environmental Editor

Pasadena Star-News
525 East Colorado
Pasadena, CA 91109
   (213) 681-4871
   Environmental Editor

Quincy Patriot-Ledger
13 Temple Street
Quincy, MA 02169
   (617) 786-7000
   Environmental Editor

Raleigh News and Observer
215 South McDowell Street
Box 191
Raleigh, NC 27602
   (919) 821-1234
   Environmental Editor

Richmond News-Leader
333 East Grace Street
Richmond, VA 23219
   (804) 649-6000
   Environmental Editor

Richmond Times-Dispatch
333 East Grace Street
Richmond, VA 23219
   (804) 649-6000
   Environmental Editor

Riverside Enterprise
14th and Orange Grove
Box 792
Riverside, CA 92502
   (714) 684-1200
   Environmental Editor

Roanoke Times
Box 2491
Roanoke, VA 24010
   (703) 981-3000
   Environmental Editor

Roanoke World-News
Box 2491
Roanoke, VA 24010
   (703) 981-3000

Environmental Editor

Rochester Democrat and Chronicle
55 Exchange Street
Rochester, NY 14614
   (716) 232-7100
   Environmental Editor

Rochester Times-Union
55 Exchange Street
Rochester, NY 14614
   (716) 232-7100
   Environmental Editor

Sacramento Bee
21st and Q Streets
Box 15779
Sacramento, CA 95813
   (916) 446-9211
   Environmental Editor

Sacramento Union
301 Capitol Mall
Box 2711
Sacramento, CA 95812
   (916) 442-7811
   Environmental Editor

Saint Paul Dispatch
55 East Fourth Street
Saint Paul, MN 55101
   (612) 222-5011
   Environmental Editor

Saint Paul Pioneer Press
55 East Fourth Street
Saint Paul, MN 55101
   (612) 222-5011
   Environmental Editor

San Bernardino Sun-Telegram
399 D Street
San Bernardino, CA 92401
   (714) 889-9666
   Environmental Editor

San Diego Tribune
350 Camino de la Reina
San Diego, CA 92108
   (714) 299-3131
   Environmental Editor

San Diego Union
350 Camino de la Reina
San Diego, CA 92108
   (714) 299-3131
   Environmental Editor

San Francisco Chronicle
925 Mission Street
San Francisco, CA 94103
   (415) 777-5700
   Environmental Editor

San Francisco Examiner
925 Mission Street
San Francisco, CA 94103
   (415) 777-5700
   Environmental Editor

San Gabriel Valley Tribune
Box 1259
Covina, CA 91722
   (213) 962-8811
   Environmental Editor

San Jose Mercury
750 Ridder Park Drive
San Jose, CA 95190
   (408) 289-5000
   Environmental Editor

San Jose News
750 Ridder Park Drive
San Jose, CA 95190
   (408) 289-5000
   Environmental Editor

Santa Ana Register
625 North Grand
Box 11626
Santa Ana, CA 92711
   (714) 835-1234
   Environmental Editor

Santa Rosa Press-Democrat
427 Mendocino Avenue
P.O. Box 569
Santa Rosa, CA 95401
   (707) 546-2020
   Environmental Editor

Schenectady Gazette
332 State Street

Schenectady, NY 12301
  (518) 374-4141
  Environmental Editor

Seattle Post-Intelligencer
Sixth and Wall Streets
Seattle, WA 98121
  (206) 628-8233
  Environmental Editor

Seattle Times
Fairview Avenue North and John Street
Seattle, WA 98111
  (206) 464-2111
  Environmental Editor

South Bend Tribune
225 West Colfax
South Bend, IN 46626
  (219) 233-6161
  Environmental Editor

Spokane Chronicle
926 Sprague Avenue, West
Spokane, WA 99253
  (509) 455-6881
  Environmental Editor

Spokane Spokesman-Review
926 Sprague Avenue, West
Spokane, WA 99253
  (509) 455-6881
  Environmental Editor

Springfield News
P.O. Box 1131
Springfield, MA 01101
  (413) 787-2476
  Environmental Editor

Springfield Union
P.O. Box 1131
Springfield, MA 01101
  (413) 787-2476
  Environmental Editor

Staten Island Advance
950 Fingerboard Road
Staten Island, NY 10305
  (212) 981-1234
  Environmental Editor

Stockton Record
P.O. Box 900
Stockton, CA 95201
  (209) 466-2652
  Environmental Editor

Syracuse Herald-Journal
Clinton Square
Syracuse, NY 13201
  (315) 473-7765
  Environmental Editor

Syracuse Post-Standard
Clinton Square
Syracuse, NY 13201
  (315) 473-7765
  Environmental Editor

Torrance Breeze
5215 Torrance Boulevard
Torrance, CA 90509
  (213) 540-5511
  Environmental Editor

Valley News
14539 Sylvan Street
Van Nuys, CA 91401
  (213) 997-4241
  Environmental Editor

Winston-Salem Journal
Box 3159
Winston-Salem, NC 27102
  (919) 727-7211
  Environmental Editor

Wisconsin State Journal
1901 Fish Hatchery Road
Madison, WI 53713
  (608) 252-6200
  Environmental Editor

Worcester Gazette
20 Franklin Street
Worcester, MA 01613
  (617) 755-4321
  Environmental Editor

Worcester Telegram
20 Franklin Street
Worcester, MA 01613
  (617) 755-4321
  Environmental Editor

BIBLIOGRAPHY

Most of the information contained in this directory was obtained by un-published materials. This bibliography lists the few generalized directories that may aid in your search for more detailed EIS-related agencies and organizations.

Greenfield, Stanley R., ed. National Directory of Addresses and Telephone Numbers, 1980-81 Edition. New York: Bantam Books, 1979.

National Wildlife Federation. Conservation Directory, 24th Edition. Washington, D.C.: National Wildlife Federation, 1979.

U.S. Environmental Protection Agency, Office of Management and Agency Services. U.S. Directory of Environmental Sources, 3rd Edition. Stock Number EPA-840-79-010. Washington, D.C.: U.S. Environmental Protection Agency, January, 1979.

U.S. Environmental Protection Agency, Office of Public Awareness. Finding Your Way Through EPA. Washington, D.C.: U.S. Environmental Protection Agency, August, 1979.

U.S. General Services Administration, National Archives and Records Service, Office of the Federal Register. United States Governmental Manual, 1979-1980. Stock Number 022-003-00982-5. Washington, D.C.: U.S. Government Printing Office, May 1, 1979.

U.S. National Research Council. Organization and Members, 1979-1980. Washington, D.C.: U.S. National Research Council, 1979.

INDEX

INDEX (All Agencies and Organizations Listed Alphabetically)

<u>A</u>

Academy of Natural Sciences, Limnology Department, 199
Academy of Natural Sciences, Systematic Biology Division, 165
Accomac County Planning Department, 9
Accomac Northampton Planning District Commission, 9
Acoustical Society of America, 161
Adams County Planning Department (IN), 9
Adams County Planning Department (WA), 9
Adams County Planning Department (WI), 9
Air Pollution Control Association, 151
Airport Operators Council International, 253
Air Transport Association of America, 253
Aitkin County Planning Department, 9
Alabama Air Pollution Control Commission, 151
Alabama Association of County Commissions, 9
Alabama Association of Soil and Water Conservation District Supervisors, 9
Alabama Conservancy, 9
Alabama Department of Agriculture and Industries, Division of Agricultural
    Chemistry, 165
Alabama Department of Archives and History, Historic Preservation, 215
Alabama Department of Public Health, Division of Solid Waste and Vector
    Control, 223
Alabama Environmental Health Administration, Division of Public Water
    Supplies, 199
Alabama Environmental Quality Association, 9
Alabama League of Municipalities, 10
Alabama Ornithological Society, 165
Alabama Water Improvement Commission, 199
Alabama Wildlife Federation, 165
Alamance Council of Governments, 10
Alamance County Planning Department, 10
Alameda County Bar Association, 143
Alameda County Free Library, 265

<u>C</u>

D

F

Forest Farmers Association Cooperative, 169
Forest History Society, 169
Forsyth County Planning Department, 39
Forsyth County Public Library System, 268
Fort Wayne and Allen County Public Library, 268
Fort Wayne City Planning Department, 39
Fort Wayne Journal-Gazette, 280
Fort Wayne News-Sentinel, 280
Fountain County Planning Department, 39
Framingham Public Library, 268
Framingham Town Planning Department, 39
Franklin County Planning Department (IN), 39
Franklin County Planning Department (MA), 39
Franklin County Planning Department (NY), 39
Franklin County Planning Department (NC), 39
Franklin County Planning Department (VA), 39
Franklin County Planning Department (WA), 39
Frederick County Planning Department (MD), 39
Frederick County Planning Department (VA), 39
Frederick County Council of Governments, 39
Freeborn County Planning and Zoning Office, 39
Fremont City Planning Department, 39
Fresno Bee, 280
Fresno City Planning and Inspection Department, 40
Fresno County Bar Association, 144
Fresno County Free Library, 268
Fresno County Planning Department, 40
Friends of Animals, 169
Friends of the Earth, 40
Friends of the Earth, Alaska Representative, 40
Friends of the Earth, Colorado Plateua Region, 40
Friends of the Earth, Mid-Atlantic Region, 40
Friends of the Earth, Midwest Region, 40
Friends of the Earth, New England Region, 40
Friends of the Earth, Northern Great Plains Region, 40
Friends of the Earth, Northern Rockies Region, 40
Friends of the Earth, Northwest Region, 40
Friends of the Earth, Sacramento Region, 40
Friends of the Earth, Southwest Region, 40
Fullerton City Planning Department, 40
Fullerton Public Library, 268
Fulton County Planning Department (IN), 40
Fulton County Planning Department (NY), 41
Fund for Animals, 169

G

Gallatin County Planning Department, 41
Garden City Newsday, 280

I

Indiana State Board of Health, Division of Sanitary Engineering, Solid Waste
    Section, 225
Indiana State Board of Health, Division of Water Pollution Control, 225
Indiana State Board of Health, Industrial Hygiene, 162
Indiana State Library, 269
Indiana Traffic Safety and Vehicle Inspection Department, 254
Indiana University, School of Public Environmental Affairs Library, 269
Industrial Forestry Association, 171
Inglewood City Planning Department, 50
Inglewood Public Library, 269
Institute of Environmental Sciences, 50
Institute of Transportation Engineers, 254
Institute of Man and Science, 50
International Association of Fish and Wildlife Agencies, 171
International Environmental Resources Network, 50
International Society for the Protection of Animals, 171
Interstate Commission on the Potomac River Basin, 50
Interstate Sanitation Commission, 225
Inyo County Planning Department, 50
Inyo Mono Association of Government Entities, 51
Iowa Association of Soil Conservation District Commissioners, 157
Iowa Citizens for Environmental Quality, 51
Iowa County Planning Department, 51
Iowa Department of Agriculture, 171
Iowa Department of Environmental Quality, 51
Iowa Department of Environmental Quality, Air and Land Quality Management
    Division, 152
Iowa Department of Environmental Quality, Chemicals and Water Quality Divi-
    sion, 202
Iowa League of Municipalities, 51
Iowa Ornithologists' Union, 171
Iowa State Association of Counties, 51
Iowa State Historical Department, 216
Iowa Wildlife Federation, 171
Iredell County Planning Department, 51
Iron County Planning Department, 51
Irondequoit City Planning Department, 51
Isanti County Planning Department, 51
Island County Planning Department, 51
Isle of Wight County, 51
Isothermal Planning and Development Commission, 51
Itasca County Planning Department, 52
Izak Walton League of America, 52

J

Jackson County Auditor (SD), 52
Jackson County Planning Department (IN), 52
Jackson County Planning Department (MN), 52

L

<u>N</u>

Q

R

<u>X</u>

None

<u>Y</u>